More Praise for *The Fate of Food*

'Timely, positive, thought-provoking.' *The Times*

'What we grow and how we eat are going to change radically over the next few decades. In *The Fate of Food*, Amanda Little takes us on a tour of the future. The journey is scary, exciting, and, ultimately, encouraging.'

Elizabeth Kolbert, Pulitzer Prize-winning author of *The Sixth Extinction*

'The challenge we face is not just to feed a more populous world, but to do this sustainably and equitably. Amanda Little brings urgency, intrigue and crack reporting to the story of our food future. Devour this book – it's a narrative feast!'

Chef José Andrés, Nobel Peace Prize nominee

'How will we feed humanity in the era of climate change? Amanda Little tackles an immense topic with grit and optimism in this fast, fascinating read. A beautifully written triumph.'

John Kerry, former US Secretary of State

'Probably the most basic question humans ever ask is, "what's for dinner?" Amanda Little – a superb reporter – helps us imagine what the answer will be as this tough century wears on. The stories she tells with such brio are food for thought and action.'

Bill McKibben, author of *Deep Economy*

'A riveting adventure story about a dire topic, but yet it somehow brims with optimism. Little travels around the world in hot pursuit of solutions, hell-bent on hope.' Julia Louis-Dreyfus

'This is a big, important book about feeding the world – but that's not why you'll read it. You'll read *The Fate of Food* because it's compulsively readable. Amanda Little takes you around the world and shows you things you never thought you'd be interested in, but now you can't get enough. Desalination! Who knew? You'll taste fish feed with her. You'll get airsick with her. You'll meet the strange, fascinating people who are solving some of the planet's most pressing problems. And, in the end, her optimism will become your optimism. We can do this.'

Tamar Haspel, *Washington Post* columnist

'*Necessity is the mother of invention*, observed Plato. Amanda Little investigates how environmental and population pressures are spurring innovation on a grand scale – with perhaps higher stakes and longer odds than history has ever seen. This is a big, sweeping story told with heart and rigor, as ambitious as it is accessible.'

Jon Meacham, Pulitzer Prize-winning author of *American Lion*

'Perhaps the greatest challenge of our century will be providing nutritious diets to 10 billion people without destroying what is left of the biosphere. Can we do it? Yes. But Amanda Little shows us that success will look nothing like today's food system. *The Fate of Food* is spectacular. The stories are beautifully woven together and filled with curiosity, openness to new ideas, and compelling insights. This book is funny, smart, dogma-free, incredibly educational, and I think will end up being an enormously valuable contribution to the world.'

Samuel Myers, professor and principal researcher,
Harvard University Center for the Environment

The Fate *of* Food

What We'll Eat in a Bigger, Hotter, Smarter World

Amanda Little

A Oneworld Book

First published in Great Britain, the Republic of Ireland and Australia
by Oneworld Publications, 2019

This edition published by arrangement with Harmony Books, an imprint of
the Crown Publishing Group, a division of Penguin Random House LLC

This paperback edition published 2020

A CIP record for this title is available from the British Library

ISBN 978-1-78607-787-5
eISBN 978-1-78607-646-5

Photograph credits
All photographs unless otherwise noted are courtesy of the author. Page 19: U.S. Department of the Interior/U.S. Geological Survey; 32: Andy Ferguson; 39: Andy Ferguson; 45: Chris Clavet, North Carolina State University; 80: Everlyne Situma; 92: Justin Kaneps; 99: Blue River Technology; 101: Blue River Technology; 108: L. Brian Stauffer, University of Illinois at Urbana-Champaign; 114: Eric Gregory Powell; 116: Eric Gregory Powell; 121: Sean Gallagher for National Geographic Image Collection/Alamy; 125: Eric Gregory Powell; 132: Chris Sattleberger/Getty Images; 135: AeroFarms; 140: AeroFarms; 150: Knut Egil Wang/Institute; 160: Sting Ray; 165: Marine Harvest; 177: Memphis Meats; 180: Memphis Meats; 184: Liz Marshall/© 2018 Meat the Future, Inc.; 188: Impossible Foods; 194: Andrew Brown/The Kroger Co.; 203: feedbackglobal.org; 210: The Compost Company; 214: Roei Greenberg; 217: Gregory Bull/File/AP; 238: Zacharias Abubeker; 246: Jim Webb; 250: Jim Richardson; 261: Food Innovation Laboratory, U.S. Army Research Development and Engineering Center; 279: Chris Newman.

Printed and bound in Great Britain by Clays Ltd, Elcograf S.p.A.

Oneworld Publications
10 Bloomsbury Street
London WC1B 3SR
England

For my mom, Nancy,

my dad, Rufus,

and my brothers.

Contents

INTRODUCTION		*1*
CHAPTER 1	A Taste of Things to Come	*11*
CHAPTER 2	Killing Fields	*30*
CHAPTER 3	Seeds of Drought	*57*
CHAPTER 4	RoboCrop	*88*
CHAPTER 5	Sensor Sensibility	*110*
CHAPTER 6	Altitude Adjustment	*129*
CHAPTER 7	Tipping the Scales	*149*
CHAPTER 8	Meat Hooked	*169*
CHAPTER 9	Stop the Rot	*192*
CHAPTER 10	Pipe Dreams	*212*
CHAPTER 11	Desperate Measures	*229*
CHAPTER 12	Antiquity Now	*243*
CHAPTER 13	What Rough Feast	*260*
EPILOGUE	Growing Up	*279*
Notes		*289*
Acknowledgments		*321*
Index		*325*

The Fate *of* Food

Introduction

OF ALL THE places I've visited, the Pot Pie room is easily one of the most surreal: a large white-walled, cement-floored space filled with big steel machines and a funnel roughly the shape and height of a back-alley Dumpster. The machines hum and chuff as conveyor belts move materials between them, but the funnel and the substance inside it have me mesmerized: a grayish blend of freeze-dried potato chunks, carrot pieces, celery and onion slivers, peas, and whey protein. I dig my gloved hands into the pallid, pebbly stuff, sifting through it the way people rummage through piled shells at the beach. It's weightless—hundreds of gallons of vegetables with the heft of confetti. For what instantly feels like too long, I stand there digging around, searching for I'm not sure what.

The mixture slowly flows down from the funnel base through a chute to another device that weighs and divides it. The portions travel to a machine dispensing bursts of beige powder made from dehydrated milk, celery salt, powdered garlic, and chicken bouillon. The seasoned mixture is then deposited and sealed, one seven-ounce portion every few seconds, into Mylar bags along with pods of oxygen-absorbing iron, clay, and salt. The bags are labeled "Chicken-Flavored Pot Pie."

This is the first stop on a tour of the Wise Company's manufacturing facility in Salt Lake City, Utah. Leading the tour is Aaron Jackson, who at the time is the company's tall, smooth, forty-three-year-old chief executive officer. We're both suited head to toe in the factory's sanitation gear, and even with his Clark Kent hair tucked under a mobcap, Jackson is magnetic. Before Wise, he'd worked at Tyson Foods, where he sold frozen chicken nugget and cutlet products. After Wise, he would go on to become the CEO of NorQuin, a large quinoa producer on a mission to mass-market the ancient grain. The man could probably sell snowploughs in the Serengeti, and despite my unease in this strange, vaguely ominous place, Jackson has me oohing and aahing my way through the "Hearty Tortilla Soup" and "Maple Bacon Pancake Breakfast" rooms, where thousands more gold and silver Mylar pouches roll off conveyors into bins. In each, technicians in white lab coats and hairnets look like Oompa Loompas as they pull levers, toggle buttons, and examine packages for flaws. At one point, to demonstrate the bag's airtightness, a stocky technician in boots places a pouch on the floor and jumps on top of it.

The scene recalls Willy Wonka's factory in part because it's achieving Wonkian ends. As a kid, I'd spent hours imagining the sensations of Roald Dahl's three-course chewing gum "made of tomato soup, roast beef and baked potato, and blueberry pie." This is a similar attempt to create an all-in-one meal that bears little resemblance to the foods it conjures—a product that when combined with a serving of hot water simulates a home-cooked dinner. "It's the food equivalent of a first-aid kit," Jackson tells me, wiping a film of beige powder from his safety glasses. "A household staple that can sustain families cut off from their normal food supply."

Jackson's products range in size from a small, $20 seventy-two-hour kit with nine freeze-dried meals to a one-year supply for a family of four that goes for $7999. Each serving is about 300 calories and costs less than $1—a per-calorie cost on par with prices at McDonald's. In his four years at Wise, from 2014 to 2018, Jackson says he more than doubled the company's annual sales, to about $75 million. The freeze-dried category as a whole had been growing during that time period, too, to about $400 million in annual sales, and that's part of why I've come to Wise—to see how real this ersatz food trend really is.

I'm skeptical going in. The survival food business smacks of zombie-apocalypse paranoia to my mind. Its success depends on the threat, real or perceived, of major food shortages in the West in the coming years. And while famine is on the rise in certain parts of the world—I'd recently seen its punishing impacts on populations in drought-stressed regions of India, Ethiopia, and elsewhere—the United States has been struggling not with a food deficit, but with caloric overload. Nearly 40 percent of our population is obese, and more than two-thirds are overweight.

The industrialized world on the whole is enjoying a more abundant, diverse, and accessible food supply than ever before in human history. The Kroger supermarket that's located a few hundred feet from my house in Nashville, Tennessee, for example, is open nineteen hours a day, seven days a week, and stocks more than fifty thousand distinct food items deriving from countries as far-flung as Taiwan and Zimbabwe. To many of us, worrying about threats against our food supply right now seems absurd, given how much there is of it.

And yet I know a growing number of people buying into the survival food trend. I'd first heard about the Wise Company from my cousin-in-law, a former cop in Zionsville, Indiana, who had stashed a supply of Wise products in his basement that could sustain his family for six months. My stepbrother, a business executive who lives in downtown Washington, D.C., has invested in a one-year supply of drinking water and long-storage food. And my brother, a climate scientist with the Nature Conservancy, has also begun building a supply in the basement of his West Virginia cabin. Part of his job is wading through reports by the Intergovernmental Panel on Climate Change (IPCC), a group of more than three thousand scientists predicting an increase in average global temperatures of at least 2 degrees Celsius by the end of this century. "I can't imagine anything worse than not being able to feed my kids," he reasons. "And the probability of major interruptions in our food supply in our lifetimes is, by almost all accounts, rising."

Granted, my brother, cousin, and stepbrother represent a skewed sample set: all are guys, all own guns, and two like to hunt in their free time with compound bows and arrows. Each possesses at least a flicker of the fatalist "prepper" sensibility that the Wise Company was founded in 2008 to serve. "Early on, our market was mostly the people preparing their bunkers for Armageddon or resisting a government they feared would take away their guns," says Jackson. Like many survival food companies, Wise was founded in Utah to serve the Mormon community, which is encouraged by the Church of Jesus Christ of Latter-Day Saints to prepare for the end of times. But Mormons and for that matter male preppers, Jackson tells me, no longer represent the entirety, or even the large majority, of Wise's exploding market.

Jackson himself is not Mormon. He grew up in the L.A. suburbs and looks more Brooks Brothers than *Duck Dynasty*, dressed under his white safety coat in a quilted jacket, ironed dress trousers, and shiny chestnut leather shoes. When he came on as CEO, he began building out the company's product portfolio to what is today dozens of different freeze-dried comfort-food products and branching out to new markets, including the camping and wilderness-adventure set, the Department of Defense, and international militaries. Eventually, he picked up distribution at Sam's Club, at Walmart, and at what is now the company's largest distributor, Home Shopping Network. "Five years ago, our market was more than 95 percent men. Today, we're reaching about 50 percent women," Jackson says, "most of them moms—guardian moms, we call them—worried about a stable food supply for their kids."

Male or female, Wise Company customers share growing fears about political and environmental instability. The first wave of customers a decade ago were concerned about inflation, economic collapse, and terrorist attacks; today, though, the major driver is natural disaster. After Hurricanes Harvey, Irma, and Maria in September 2017, the Federal Emergency Management Agency sourced about two million servings from Jackson for their relief effort. "It's not just the freak events. We get calls from people saying, 'I live in Miami and flooding is now routine. I'm worried Florida is going to be under water in two years,'" Jackson tells me. "Or from people in upstate New York who experienced a one-in-a-thousand-year blizzard and couldn't get out of their driveway for two weeks." Calls from people who saw the impacts of Hurricanes Katrina and Sandy, who lived through the 2014 drought and 2018 forest fires that scorched California, and know maybe the government won't come to their rescue when a disaster hits. "They're adopting the attitude that, well, luck favors the prepared."

I have yet to invest in this luck, in part because I'm optimistic enough to believe that I won't need to. But Jackson made a compelling case that the survival food industry is as much a function of pragmatism as it is of paranoia among a growing number of people who realize they're up against increasing environmental threats on the one hand and diminishing government safety nets on the other. It's not just happening in the United States—every country in the world today is facing environmental volatility, and many are also politically unstable. As the United Kingdom departs the European Union, there may be at least some disruption to its food supply, especially in a no-deal scenario, which could reduce the availability and increase the cost of some products, including fresh fruit and vegetables. There won't be serious near-term shortages, but down the line, the threats are real. Ultimately, you're left wondering, how screwed *are* we exactly?

BEFORE WE CONSIDER the extent of the current threats to our food supply and, in particular, the perils of modern farming, let's quickly review some of the achievements of industrial agriculture. As many as two billion people might not exist if it hadn't been for the advent of agribusiness. Farms globally now produce 17 percent more calories per person than they did in 1990. And while some 800 million people still suffer from chronic hunger, that is almost 200 million fewer than there were thirty years ago. Meanwhile, prices have fallen. The average household in the 1950s spent about 30 percent of its budget on food. Today, we spend about 13 percent—a financial advantage for low- and middle-income households, and a boon for economies worldwide. Processed foods have also liberated men and, in particular, women from the drudgery of preparing every meal from scratch. Yet the disadvantages of abundant, low-cost food are well documented, starting with massive waste, overconsumption, poorer nutrition, and a reliance on fewer, more concentrated farms to feed the world. There's also an increasing risk that the methods we've devised to feed billions more people are backfiring on the environment.

By the time I joined Jackson for the Wise tour, I'd traveled to thirteen states and eleven countries researching the changes, both subtle and radical, taking place in our food system. By "food system," I mean the vast network of local and international growers, processors, and distributors who feed seven and a half billion people worldwide. I wanted to understand the effects of population growth and climate change on agriculture in fast-growing countries, including China, India, and throughout sub-Saharan Africa. In March 2014, the IPCC reported that droughts, flooding, invasive species, and increasing weather volatility were already hurting agricultural productivity worldwide, and that permanent drought would become the norm in large portions of most populous nations by midcentury, including throughout the American Southwest—a swath of highly populated land stretching from Kansas to California and down into Mexico. The IPCC projections are flat-out scary: they show that current warming trends could cut global crop yields 2 to 6 percent every decade going forward—that's millions of acres of farmland phasing out worldwide every ten years—even as the global population climbs.

In October 2018, the IPCC released a sequel report concluding that at the current rate of emissions, the atmosphere will warm up by as much as 1.5 degrees Celsius above preindustrial levels by 2040, a fate that would radically transform our living conditions. "It's like a deafening, piercing smoke alarm going off in the kitchen," said Erik Solheim, executive director of the UN Environment Programme (UNEP). "We have to put out the fire."

According to Jerry Hatfield, the director of the U.S. Department of Agriculture's National Laboratory for Agriculture and the Environment, the single biggest threat of climate change is the collapse of food systems: "Other threats—flooding, storms, forest fires—may be more sudden and severe in certain regions, but disruptions in food supply will affect virtually everyone," he tells me. Tim Gore, the head of food policy and climate change for Oxfam, puts it this way: "The main way that most people will experience climate change is through its impact on food—what they eat, how it's grown, the price they pay for it, and the availability and choice they have." Joyce Msuya, UNEP's deputy executive director, cautions that the least wealthy nations are the most vulnerable: "The agriculture sector is quite dominant in most of the world's developing countries. Here, there's a dichotomy of huge demand—more mouths to feed—while environmental pressures are shrinking the food supply."

Food prices, according to the IPCC, could nearly double by 2050 given current climate and population-growth trends. If they do, conflicts over limited affordable food would probably escalate and further imperil global food security—a scenario that might force my brother to tap those emergency food provisions he's been stockpiling, if he hadn't exhausted them already. International conflicts over food resources could interrupt trade and paralyze distribution networks, and given that the United States imports more than half of its fruit supply and about a third of its vegetables, the result would be, among other things, lots of empty shelves at your local supermarket. The shortages may be particularly acute in countries like the U.K., where half of the entire food supply is imported.

Little wonder that there are other "post-food" companies betting on disruptions. The Silicon Valley–based start-up Soylent Inc., with about $70 million in funding, has produced a kind of adult baby formula—a vegan beverage designed to replace a nutritionally complete meal, saving consumers time and money while reducing their carbon footprint. Soylent has been so popular that it's given rise to Super Body Fuel, Ample, Koia, and half a dozen other new meal-replacement brands. Meanwhile, the Pentagon's research division is developing sustenance for soldiers that can be cranked out on demand by portable 3-D printers. Sensors on the soldiers' bodies will detect, say, a potassium or vitamin A deficit and send that data to the 3-D printer, which will then generate customized, nutrient-fortified food bars and pellets from flavored liquids and powders. The technology is expected to be in the field by 2025; it's a future that most of us can't imagine inhabiting.

AFTER MY VISIT to the Wise factory, I whip up a bowl of rehydrated pot pie. In truth, I ask my kids to do it. They fire up the electric kettle, pour, stir, wait for the pebbly chunks to soften. To them, it's a simple science experiment. To me, it's confronting a future I don't want to meet.

But Jackson's core technology is not new. The Wise Company practices a twenty-first-century version of something the Incas started back in roughly AD 1200 when they placed potatoes and *ch'arki*, a kind of beef jerky, on elevated stone slabs to freeze overnight and then quick-dry in the sun. It wasn't until World War II that modern freeze-drying methods were developed to preserve blood serum for wounded soldiers. The current processes arose in the late 1970s, when concerns over the oil crisis and stagflation motivated millions of Americans to cache food. The Wise Company has tweaked this decades-old formula only a little: fresh ingredients are rapidly "blast-frozen" at temperatures as low as negative 80 degrees Celsius to prevent the formation of ice crystals that could affect food texture and nutrition. The food is then placed in a heated vacuum chamber that causes the ice to "sublime," changing directly from a solid to a gas without passing through a liquid phase. Pores left from the vanished ice quickly absorb water when the foods are rehydrated. The process takes nearly double the energy used for canning but retains about 90 percent of the food's nutrients and preserves it for far longer. Wise products, which guarantee a shelf life of twenty-five years, can conceivably remain edible, according to Jackson, for "gosh, centuries."

The pot pie mixture looks nothing like the real thing when we serve it up. It's a tawny gruel. I hesitate, stifle a gag reflex, channel my inner Violet Beauregarde, and swallow. The stuff goes down easy, tasting like my grandmother's chicken casserole. But when I imagine a world in which my grown children are surviving in our basement on Mylar pouch meals as they struggle to rig Mark Watney–style indoor cropping systems, I lose my appetite and wonder: What will be on the table when

I'm visiting my grandkids for Thanksgiving dinner in the year 2050? Will future historians look back on our current agricultural moment and see it as Dickens did Europe in the late eighteenth century—an age of belief and incredulity when "we had everything before us, we had nothing before us"?

There are certain passages in the IPCC report that seem to indicate we're headed toward nothing. By the middle of this century, the report reads, the world may reach "a threshold of global warming beyond which current agricultural practices can no longer support large human civilizations." That fate hinges on a key assumption though—that current agricultural practices won't change. And if my travels have taught me anything, it's that farmers, scientists, activists, and engineers the world over are radically rethinking food production.

Environmentalist Paul Hawken, editor of *Drawdown: The Most Comprehensive Plan Ever Proposed to Reverse Global Warming*, found that eight of the top twenty most effective solutions proposed by his team of scientists were in the realm of agriculture: "Among one hundred strategies we researched across all categories of society and industry, the food solutions are the most curative and impactful."

It's hard to overstate how much the global food system has changed in the last thirty years, and harder still to know how and how much it will change in the decades ahead. In *The Fate of Food* I investigate what that change might actually look like. I, like most of us, love food too much to accept a future of freeze-dried chicken pot pie. ("One cannot think well, love well, sleep well, if one has not dined well," wrote Virginia Woolf.) And I've found good reasons to expect that it's a future we can circumvent. Innovation and ignorance got us into the mess we've made of our food system, and innovation combined with good judgment can get us out of it.

In the following chapters, I'll explore whether and how we'll feed a hotter, drier, more populous world sustainably and equitably—and with far more on the menu than rehydrated comfort foods. I'll meet people like Jorge Heraud, a Peruvian-born engineer who builds robots that can weed crops, cutting the use of agricultural chemicals. I'll visit

Vat of vegetables

start-ups creating lab-grown and plant-based meats. I'll travel into the fields of Kenyan farmers planting the country's first GMO maize, and to the world's largest vertical farm, where vegetables are grown without soil or sun. I'll venture inside the smart water networks of Israel and the world's largest fish farm in Norway. And I'll meet people renewing old ideas, like practitioners of permaculture farming and harvesters of edible insects and botanists reviving ancient plants.

Along the way, I'll find some answers to the questions I later realized I'd been digging for in that funnel full of freeze-dried vegetable chunks—not just about all the trouble we're in, but about how we'll get out of it.

CHAPTER 1

A Taste of Things to Come

Plenty sits still, hunger is a wanderer.
—SOUTH AFRICAN PROVERB

THIS BOOK GREW out of some seeds I planted in my backyard garden in April 2013. The diehards among you may recognize that previous sentence, which echoes a line in the introduction to Michael Pollan's classic *The Botany of Desire,* a book that inspired garden lust in me and many other readers. But my seeds and Pollan's yielded different results. His bourgeoned; mine bombed.

First, they sprouted and, as seeds miraculously do, grew to be fifty and a hundred times their original size. They produced glossy green leaves and the advanced stages of fruit before things took a turn. The garden had been partly my idea, partly my kids'. They'd been raising herbs and cherry tomatoes in planter boxes on their school playground, and wanted to try it on a larger scale at home. It didn't take much convincing—I'd already read not just Pollan but also Mark Bittman, Dan Barber, Alice Waters, et al. I'd drunk down the all-natural, agave-sweetened, anti-Kool-Aid Kool-Aid. Tending our own backyard plot would slow down the rhythms of life a bit, my husband and I agreed, improve our daily vegetable intake, and offer an iPad antidote while creating a deeper bond with nature. And hadn't I read somewhere that bonding with nature helped kids focus and improved hypothalamus activity? And family unity! This would be a good, healthy, mind-expanding, family-unifying, money-saving, world-bettering weekend activity.

Nashville, Tennessee, is a far cry from Berkeley, California, but our community teems with acts of world-bettering, especially among parents and especially in relation to food. We have expanding populations of backyard farmers and vegans and Paleolithic dieters and people who raise their own chickens. I have good friends who would rent an ox and a Mesopotamian plough to get closer to the ancient roots of their nourishment if they could, so deep is their food nostalgia. I am nowhere near this serious about my own family's diet. I love a good farmers' market, but I mostly shop at my local Kroger and have no qualms about feeding my kids out-of-season fruit or, for that matter, the school lunch. We buy organic when we're feeling flush, which means we often don't, and we go through at least half a dozen mass-produced apples a week—the kind of fruit dismissed by critics as overengineered sugar bombs. Still, I'm prone to nostalgia for a time before industrial agribusiness, for the bygone era of heirloom flavors that farmers' markets and backyard gardeners are trying to protect.

So that spring of 2013, we went all in, sinking hundreds of dollars into a fenced-in ten-by-fourteen-foot raised bed, a small mountain of compost, tomato cages, fish-oil fertilizers, and crates of organic and heirloom seedlings, only to discover that I'm the gardening equivalent of tone-deaf. Two months after planting day, I stood inside my chicken-wire fence, scanning the wilting husks of half a dozen once promising cornstalks, a patch of elephantine cucumbers, one the width of a raccoon, and an aphid infestation in five tomato plants that had merged into a unified organism. The evidence was clear that I'm no better suited to growing my family's food than I am to repairing a circuit board. An idea that had originally seemed eminently practical—a twenty-first-century victory garden—proved, for our family at least, to be not very practical at all.

It isn't basic knowledge that I lack, it's time, vigilance, and good

My overgrown garden

judgment. I have some unique handicaps, admittedly. Pruning edible plants feels to me like a mild form of infanticide—I avoid it, along with slugs, mites, aphids, and stinkbugs, and the application of whatever organic pesticides might deter them. The mosquitoes get so bad in our backyard that they, combined with the seething summer heat of Middle Tennessee, often dissuade me from watering and weeding. And when I do get up the courage to tackle weeds, I often can't distinguish them from the seedlings and let them grow.

Even now, many years after that first failed attempt at edible gardening, we're still trying to raise vegetables in our backyard plot. The results have improved a bit with help from my husband and kids, who have become more reliable farmhands. But if I'm honest, we haven't produced much in the way of reliable or abundant dividends. The garden ultimately costs us more than it saves. We keep at it because it makes us feel good. It engages the senses, connects us to the land we live on, and looks nice at a safe distance. The presence of it calms my concerns about, if not a secure food supply, then about the broader impacts of technology on our lives.

And as it turns out, that first garden did manage to be generative—not of food but of questions, such as: How will we fix a failing food system if we can't necessarily rely on a critical mass of enlightened, vegetarian, non-GMO, organic-only, backyard-harvesting consumers to do so from the ground up? It also got me exploring the history of agriculture and the technologies that have transformed it along the way. I learned that the food-growing efforts of individual producers have been riddled with hardships and impracticalities since, well, the beginning of human civilization.

THE YEAR IS 4000 BC, and we're not too far from present-day Baghdad. A Mesopotamian farmer is growing wheat on a farm located somewhere between the Tigris and Euphrates Rivers. He, or maybe she, is hitching an animal to a tool that looks a lot like a hoe but is really a protoplough. At this point, humans are about six thousand years into food cultivation, and there's no consensus theory (six millennia later, there still isn't) to explain how and why we made the move as a species from plant-gathering to plant-taming. But if you asked her, this wheat farmer might tell you it's simple—because her family liked staying put. (Or maybe her kids, like mine, came home one day wanting to plant what they'd seen growing.)

There's little dispute among archaeologists that farming made it possible for settlements, and ultimately for civilizations, to thrive over time. But there's also evidence that many settlements predated farming. There were religious sites with temples and permanent dwellings established long before the first cultivated crops appeared. Places such as Pikimachay in western Peru and Gobekli Tepe in eastern Turkey were located, circa 10,000 BC, near fishable rivers or in regions where the food supply was easy pickings. Wild sources of grains, fruits, and protein were abundant and reliable—until they weren't. A drought or blight may have come along, or the populations outgrew the wild food supply, and the settlers had to find ways to make do with whatever edible plants remained.

We'll probably never know who pushed the first seeds into the soil and tended those original harvests, or exactly why, but it's clear that by the time we got to prehistoric Mesopotamia, humans had by and large decided that it's better to grow than to gather what you need. We stopped wandering the natural world and began to shape it. Migratory lifestyles gave way to settled societies. Ancient economies began to form. Fertility rates shot up, and populations expanded. Larger families were easier to care for when you weren't constantly on the move, and more offspring meant extra hands in the fields.

In his book *Sapiens: A Brief History of Humankind,* historian Yuval Noah Harari writes, "We did not domesticate wheat. It domesticated us. The word 'domesticate' comes from the Latin *domus,* which means 'house.' Who's the one living in a house? Not the wheat. It's the sapiens."

But as houses were built and populations soared, nutrition declined. Farming radically narrowed the diversity of available foods. Foragers had subsisted on a varied, protein-rich diet, but now farmers were living off whatever limited monocultures of grain they could produce. Bioarchaeologists have found lesions on the skulls of the settlers in early farming societies, indicating severe iron deficiencies, along with evidence of stunting from poor nutrition. The first farming populations were almost invariably shorter than their hunter-gatherer predecessors and more vulnerable to disease. They dealt with longer and more grueling workdays. Farming required land clearing, hoeing, planting, weeding, warding off pests, harvesting, storing, and distributing the food—more calorie-intensive labor than gathering wild bounty.

"Drudgery and hunger provided a motive for developing tools," says Columbia University professor and ecologist Ruth DeFries. "Every new agricultural tool introduced since the first farming settlements has been designed with the same goal: to coax more food from the earth with less human effort." This is useful context as we consider how we'll feed a hotter, more populous world in the coming decades. Humans have now spent the better part of ten thousand years developing a succession of tools to this end, all of them temporary solutions that, generation after generation, get replaced or upgraded to work on larger scales.

We dammed streams first, then rivers. We constructed hand tools from stones and wood and then metals and eventually supplanted those tools with machines. We made fertilizers from human waste and animal manure and then from complex chemicals. Now we have sensors and robots to interpret the needs of our crops; we have food that can be grown without sun or soil; we have Mylar-packaged meal replacements. "Each agricultural technology has been one more link in the lengthy chain of experiments aimed at producing a bigger, more reliable food supply with less work," says DeFries. Exploring this chain of experiments is a recurring theme of this book. In each chapter, I try to understand not just where we're going but how we got here, having moved through the long-running technological continuum of food production.

The first Mesopotamian farmer to rig his plough to an ox was an early link in the chain. He'd found a way to tap into the power of animals. Tilling soil took him a fraction of the time and energy it had taken using human power alone. Later generations of Mesopotamians would learn to attach a mechanism to the plough that fed seeds into the soil as it was turned, automating the planting process and increasing yields.

As farmers began to produce crops in volumes well beyond the needs of their communities, they became merchants. Advances in food preservation and storage—sealed containers, drying, fermenting, and curing—meant that food could travel farther afield. The Iron Age brought larger and sturdier ships and trade routes extended across oceans. Emerging empires and dynasties—Spartan, Roman, Zhou—began to specialize in different food exports: grains, nuts, spices, oils, fruit, wine, salted meats, and dried fish.

By AD 700, Muslim traders had established the early foundations of the global economy, distributing crops from northern Africa, China, and India throughout Islamic lands. Imports meant more diverse and nutrient-rich diets and better health. Merchants also traded their ideas and beliefs along with their provisions. The Prophet Muhammad, founder of Islam, was a spice trader when he started preaching, and for more than a thousand years, his Muslim disciples distributed the Koran along their trade routes as they sold their coveted cinnamon, cloves, nutmeg, and peppercorns.

For all that we don't know about the human transition from hunting and gathering to "Would you like a Koran with that?" we can safely assume that agriculture was not a fluky discovery or happy accident, but a gradual, often painstaking process that arose from choice or necessity. We can assume that the benefits of farming—a controllable food supply, a lower risk of starvation, and the comforts of staying put—eventually outweighed the costs. Food surpluses meant that economies could diversify. People could choose *not* to farm and do whatever else needed doing—designing tools, constructing homes, creating art. Students could learn, builders could build, governing bodies could form in societies no longer roaming in search of food. Neolithic farming settlements gave rise to the first written languages, to ceramics and glass production, to irrigation and wheeled transportation systems, and eventually to a mastery of metals and machines.

Over time, robust food systems conferred political power. The Bible bears this out in the Old Testament story of Joseph, who interprets the dreams of his prison guards when he is locked in an Egyptian dungeon. The pharaoh summons him after two haunting dreams—first, that seven sickly cows eat seven healthy cows, and then that seven thin heads of grain swallow seven fat heads. Joseph tells the pharaoh that seven years of famine in Egypt will follow seven years of abundance. The king prepares accordingly, stockpiling grains during the productive years. Sure enough, the subsequent famine is so far-reaching that people come in droves to Egypt from across the world to buy grain. Pharaoh gives Joseph fine robes and the keys to the kingdom.

For thousands of years, civilizations from the Mayans of Mesoamerica to the Vikings of Scandinavia rose as their food supplies flourished and fell as they declined. Even today, the nations with the least reliable food supplies generally have the least diverse economies and the most vulnerable governments. In 2014, for example, the Pentagon warned that drought and crop failures throughout the Middle East—the region that once comprised the Fertile Crescent—empowered ISIS and other extremists to recruit followers among starving and displaced populations. Just before that, in 2011, hunger had helped foment the

Arab Spring after droughts had crippled wheat fields in Russia and the United States, causing prices to spike worldwide.

We can only expect these trends to intensify in the coming decades: the countries and communities that most creatively address their food supply challenges will be the ones that are best equipped to succeed.

THE FIRST MAJOR panic over global food supplies began to percolate in the late 1700s. Arable land was declining as urban populations rose. The English parson Thomas Malthus announced in 1798 that food supplies could not keep pace with demand: "The power of population is so superior to the power in the earth to produce subsistence for man that . . . premature death must in some shape or other visit the human race." This theory went mostly ignored until the mid-1840s, when famine swept through Ireland. But then the unexpected occurred, a scientific serendipity: chemists discovered that nitrogen and phosphorus, which had been stripped by overfarming from European soils, were the essential life-giving elements of plants. Within a few decades, the German chemist Fritz Haber had cracked apart a molecule of atmospheric nitrogen, producing the main ingredient for the world's first synthetic fertilizers.

Malthus hadn't foreseen the coming era of chemicals or of mechanization. The first reaping machines arrived on the market in the mid-1800s, followed by the first steel plough, and by 1903 an American factory was producing combustion-engine tractors. Work that had once required many days of human and animal labor to complete now took just a few hours. Crop breeding underwent a similarly radical transformation around the same time. In 1856, the Austrian monk Gregor Mendel began his famous experiments in his monastery's garden studying heredity in peas. The following decade, Charles Darwin published his book on cross-fertilization in plants. It wasn't long before American scientists applied Mendel's findings and Darwin's theories to their quest to breed better corn and wheat. They isolated and combined specific traits to produce faster-growing, higher-yielding, pest-resistant crops. The invention of hybrid seeds combined with the arrival of chemical pesticides and fertilizers brought on the paradigm shift known as the Green Revolution.

What ensued was a productivity explosion, an agricultural H-bomb. In the five decades after World War II, the global food supply jumped 200 percent. The world's population more than doubled, in turn. Factory farms subsumed family farms and crops began to derive their energy from fossil fuels. Agribusiness could now produce immense quantities of wheat, soy, and, in particular, corn, which was then processed into products ranging from corn syrup and maltodextrin (a food additive) to, most notably, meat. A full-grown, 1200-pound steer consumes thousands of pounds of corn and soy feed throughout its lifetime, and produces less than half its total weight—500 pounds or so—in edible beef.

There are plenty of upsides to the Green Revolution. Industrial farms have addressed many of the inefficiencies and impracticalities of individual and small-scale food production. Journalist Paul Roberts writes that the modern food system has been "celebrated as a monument to humanity's greatest triumph." He adds that by the late twentieth century we were "producing more food—more grain, more meat, more fruits and vegetables—than ever before, more cheaply than ever before, and with a degree of variety, safety, quality and convenience

A dam transforms an agricultural landscape

that preceding generations would have found bewildering." Economies broadly have prospered from more abundant and affordable foods. But there are the consequences to consider: for every gain in food production, there has always been a cost.

No Neolithic farmers could have foreseen the impacts of the juggernaut they'd set in motion. Who could have imagined that planting stray seeds of einkorn wheat might result in a practice that over the course of twelve millennia would transform nearly half of the world's habitable land? Farming has altered the natural systems of the earth more than any other single human activity. Virtually every major river on earth has been tapped or dammed, every major lake and aquifer exploited, and most (about 70 percent) of that freshwater flows to farms. Fisheries now harvest more than a third of the edible biomass from the ocean's coastal waters. Agribusinesses in just the last couple of decades have consumed areas of bio-rich forestland collectively the size of Peru. Livestock operations globally produce nearly five billion cattle, pigs, goats, and sheep, up 25 percent in three decades. Together those animals graze an expanse of land larger than the African continent.

The architects of the Green Revolution had a grand goal: ending hunger worldwide. Norman Borlaug, the father of hybrid wheat, said when accepting the 1970 Nobel Peace Prize that he hoped "to provide food . . . for the benefit of all mankind." He didn't. Modern agriculture now produces many more calories per person than it did at the end of World War II—about eight hundred calories more for every person on earth, if everybody got the same portion. They don't, not by a long shot. The nutrient gap between rich and poor populations has grown wider in the past half century—which is to say, the rich have far more nourishment. "It's tempting to idealize the small-scale farming systems in the developing countries as uncorrupted by modern tools. In reality, the yields are low, farmers are facing high risks and heavy debts, daily life centers on staving off hunger, and hunger persists on a grand scale," Ruth DeFries cautions. More than 800 million people today are undernourished, and the lopsided distribution of calories on our planet remains one of the Green Revolution's great failures. Cheap food and long, inefficient supply chains have also led to a waste pandemic. Roughly a third of the food produced worldwide rots in transit or is thrown out.

There are also the unintended consequences of farm chemicals to consider. Excess fertilizers applied to farmland run off into lakes and oceans, causing algae blooms that suffocate aquatic life. Herbicides and fungicides have squelched vital bacterial activity in the microbiomes of topsoil. Pesticides have caused mass die-offs of bees, beetles, and butterflies that play Oscar-caliber supporting roles as pollinators in food production. Monocultures, or fields composed of a single crop, have been grown for millennia—since those early fields of einkorn—and they're cheap, fast, and efficient to plant and harvest using machines, but they're gravely inefficient when it comes to pests. A vast stretch of a single crop is an all-you-can-eat buffet for certain insects and fungi. Those pests are evolutionarily adept at developing resistance to chemicals, requiring more and stronger chemicals in turn. This is one reason why, in the forty years between 1960 and 2000, pesticide use in the United States doubled.

Huge investments of fossil fuels go into the production of agrochemicals, and into the machines and transportation networks that produce and distribute food—all of it adding up to a monster carbon footprint. The single biggest blowback of the Green Revolution is climate change. Absurdly, the greenhouse gases that now threaten the future of the world's farms are also largely produced by the farms themselves, especially the big mechanized ones. Most of us generate more planet-warming emissions from eating than we do from driving or flying. Food production now accounts for about a fifth of total greenhouse gas emissions annually, which means that agriculture contributes more than any other sector, including energy and transportation, to climate change.

Ruth DeFries points out that for the better part of ten thousand years, the problems of human food production have derived from scarcity—too little fertilizer, arable land, or energy. But now, many of the problems come from overabundance—too many chemicals, too much CO_2, too much waste. The downsides of all this abundance go well beyond the environmental costs. Higher yields have led to a decline in nutrition. Government studies have shown dropping levels of protein, calcium, potassium, iron, and vitamins C and B_2 in dozens of vegetable and fruit crops over the last fifty years. During that time frame, the mass marketing of highly processed foods has driven consumers—Americans, in particular—toward foods high in calories but low in nutrients. Average sugar consumption in the United States has jumped more than 20 percent in three decades, and during that time the weight of the average American adult has increased about 20 percent. The prevalence of diabetes has increased 700 percent.

The Green Revolution, for all its advantages, created a food system that does not nourish people equitably—one in which some populations are severely overfed, others severely undernourished, and still others are at the same time overfed and undernourished. This last category is the fastest growing: nearly half of all countries worldwide are now experiencing serious levels of both undernutrition and obesity.

ON ITS OWN, the failure of my backyard garden did not make me feel helpless to change the food system. It was one factor among many, including the times I had tried and failed to be a vegan and then a vegetarian and then a pescatarian and then a responsible carnivore who eats only local, humanely raised meats; tried and failed to eat seasonally and all-organic and to cut out GMOs. I want to grow and buy and prepare the right leafy, fresh-picked things, for my kids especially, but I work full-time and cook last minute and have a weakness for processed snack foods. I also have an appetite for burgers, brisket, fried chicken, roast turkey, every kind of breakfast meat, and—it's practically mandatory for residents of Tennessee—barbecued ribs.

My meat habit, in particular, plagues my conscience. I'm well versed in the cruelty of conventional livestock operations and in the climate impacts of my choices, and I'm aware that no self-respecting environmentalist can ignore that. I know a serving of beef has a carbon footprint that's four times greater than a serving of chicken, which in turn has a carbon footprint three times greater than a serving of lentils. I've gone for months without meat—until I end up back at the barbecue.

Sustainable food advocates have scrupulously examined the flaws in our food system, but the large-scale solutions they've explored, if they've explored them at all, are relevant mostly to people who have the time, income, and creativity to cook from organic produce in a vegetable box delivered direct to their door. They often suggest that we dismantle the industrial farms that produce the majority of U.S. crops, and that we reject genetically modified seeds, which are used for more than 70 percent of American corn, soy, cotton, and rice production. They say we should adjust, in turn, to significantly higher food costs. Some price increases may be inevitable in the coming decades, but steep jumps would strain most consumers. As food historian Bee Wilson puts it, "No one has yet discovered how to raise prices for the overfed rich without squeezing the underfed poor."

High-priced food is fetishized by our current culture of haute gastronomy, which celebrates, for example, the Octopop, a "low temperature cooked octopus dipped in orange and saffron carrageenan gel and suspended on dill flower stalks," as described by celebrity chef Adam Melonas. Popular shows like Netflix's *Chef's Table* and PBS's *The Mind of a Chef* feature thousand-dollar servings of "vintage côte de boeuf" and hundred-dollar confections covered in edible gold leaf, drawing millions of viewers who are likely couch-bound and snacking on some of the lowest-quality, most overprocessed sustenance in human history. Even as a meager cook with a limited food budget, I buy into the collective fantasy life fed by this media. I'll see a recipe in, say, *Saveur* magazine, for Braised Zabuton with Espresso Rub, and I get a reflexive urge to run to Whole Foods, even though I have no idea what a zabuton is. Then I remember the strange paradox that I'm participating in a culinary culture of excess at a time when the modern food system has perhaps never been more imperiled. I get the nagging sense that I'm fiddling with zabuton while our planet begins to burn.

THERE'S A DEEP distrust of technology as applied to food. That's partly because the food industry has blundered its way through so many costly failures. DDT, for example, a chemical pesticide developed in the 1940s, was applied to crops for decades before scientists figured out it was killing birds and causing a fourfold increase in breast cancer. It's one of several agricultural chemicals approved by the United States government and later banned, long after people had been sickened and ecosystems harmed. Saccharin and aspartame, to take another example, were promoted as innovative low-calorie alternatives to sugar before they were exposed as carcinogenic to rats. Margarine was likewise sold as a shelf-stable, heart-healthy alternative to butter before it came to light that the stuff contained heart-harming trans fats. And the many food ingredients derived from millions of acres of U.S. corn, from maltodextrin to monosodium glutamate, while innovative and profitable, have not been advantageous to human health. In these cases and countless others, technology backfired. It made our food system not smarter but more flawed.

Justifiable concerns have been raised, too, about the rampant use of herbicides (the weed-killer Roundup was deemed a threat to human health by the World Health Organization (WHO) in 2015, after forty-one years on the market) and synthetic food coloring (now linked to hyperactivity disorder in kids) and salmon farming (which fattens fish on corn feed, a substance that has never belonged in any natural aquatic system) and genetic engineering of crops (a method that "has fallen short of the promise," according to page one of the *New York Times*).

It's not surprising that few American consumers are applauding the onset of the Next Green Revolution, as *National Geographic, Wired,* and others have dubbed the new wave of agricultural technologies now in development. "Food is ripe for reinvention," Bill Gates proclaimed in 2014 at a Microsoft shareholder meeting. Huge flows of public and private investment—including billions from companies inside the conventional ag industry and outside of it, like Microsoft, Google, and IBM—are now funding new methods of food production. A generation of entrepreneurs in fields as varied as plant genetics, aquaponics, big data, and artificial intelligence are vying to build a better, "smarter," more resilient food system and to create new ways to harness the sustenance it yields.

Some are in it for do-gooder reasons—to feed the world sustainably and equitably while curbing climate change—while others see a gold mine: nine billion mouths to feed. They know that whoever cracks global food security will, like Joseph, get keys to the kingdom when the lean days inevitably come. Whatever the motivation, there's a controversy building. Most sustainable food advocates bristle at the idea of reinventing food—they want it *de*invented, thank you very much. They advocate for a return to preindustrial, pre–Green Revolution, organic, and biodynamic farming practices to which skeptics inevitably respond, "Yes, that's nice. But does it *scale*?" Sure, a return to traditional methods might produce better food, but can it produce *enough* food?

The rift between the reinvention camp and the deinvention camp has existed since Norman Borlaug began breeding modern wheat, setting in motion a dispute that is now a raging battle of hyperbole, reactionism, and platitudes. One side views technology as corrosive, the other sees it as a panacea. One side covets the past, the other the future. In his recent book, Charles Mann calls the reinvention camp "Wizards" and the deinventionists "Prophets": "Wizards, following Borlaug's model, unveil technological fixes," Mann writes. "Prophets . . . decry the consequences of our heedlessness." He elaborates: "High-yield, Borlaug-style industrial farming, Prophets say, may pay off in the short run, but in the long run will make the day of ecological reckoning hit harder. The ruination of soil and water by heedless overuse will lead to environmental collapse, which will in turn create worldwide social convulsion. Wizards reply: *That's exactly the global humanitarian crisis we're preventing!*"

As someone observing this debate for years, I've come to see it's not serving us well at all, and to wonder: Why must it be so binary? Why can't we do some version of both? It seems to me there can—there *must*—be a synthesis of the two approaches, like the arrow out of the two-sided opposition in the classic Hegelian dialectic we learned in high school. Our challenge is to borrow from the wisdom of the ages and from our most advanced technologies to forge a kind of "third way" to food production. Such an approach would allow us to improve harvests while restoring, rather than degrading, the underlying web of life.

SEVERAL YEARS AFTER my 2013 garden failure, I meet Chris and Annie Newman, a young husband-and-wife team of farmers who live in the northern neck of Virginia. She's an artist, he's a software programmer, and together they make a very convincing case that ecological farming and technological innovation can do more than coexist, they can achieve a powerful synergy. I would meet many others throughout my research—scientists, activists, chefs, engineers, executives, programmers, educators, farmers—who are helping to model, like the Newmans, a third way.

Many of these folks appear in the coming chapters, and at the end of the book we'll visit Chris and Annie at their farm at the edge of the Potomac River, which they share with their two little kids, hundreds of chickens, dozens of hogs, and growing groves of fruit and nut trees.

I first encounter Chris Newman virtually, when I read a manifesto he's published on Medium.com, "Clean Food: If You Want to Save the World, Get Over Yourself." A few weeks later, having read that and everything else Chris has written about farming life (which is a lot), I show up on his doorstep. Chris grew up in a predominantly black neighborhood in Southeast Washington, D.C., with an African American mom and a Native American dad. He was a child whiz who later became a software programmer doing high-level work for the Treasury Department. The work was relentless, and Chris became physically sick with stomach pains. He went through months of biopsies and colonoscopies and a series of specialists before the problem was diagnosed—first by Annie, and later by the doctors—as stress.

In 2013, when trying to recover, he read a book he found in a pile lent by neighbors, Michael Pollan's *The Omnivore's Dilemma*. Chris took an interest in one of its characters, Joel Salatin, founder of Polyface Farm, who practices holistic animal husbandry, integrating diverse crops and livestock in a way that mimics natural systems. He recognized certain patterns in indigenous agriculture he'd learned as a kid, and within days had signed up for a workshop with Salatin. Soon he and Annie were hatching a plan to start farming. By 2018, after five years of selling high-end organic meats and vegetables, Chris had begun to call into question the efficacy, and even the ethics, of his own pursuit.

"I'm a permaculture farmer," begins his manifesto. "My goal is to develop natural ecosystems that produce food. My dream is a world with ready access to a diet that nourishes the body of the consumer, provides a living for the producer, and leaves the Earth joyfully habitable. I share that dream with a lot of people who call themselves permaculturalists, natural farmers, plantsmen, or foodies. I fear, however, that [we are] succumbing to tribalism; forgetting that saving the world means saving all of the people in it; even the ones that love cheap burgers and Coke. We're digging foxholes and making monsters out of people who don't agree with us, or who don't understand, or who do understand but are powerless to act."

Chris Newman inspects a chicken

Newman goes on to describe the "accessibility gap" that plagues sustainable food production, or "Clean Food," as he's dubbed it. The products he and Annie sell include $12-a-pound pork chops and $4-a-pound chicken, which only high-end markets and restaurants can afford. Bringing down their prices would mean going out of business: "Our food is not accessible. It's just not. It's beyond the wallets of damn near everybody; it's the biggest problem with sustainable food systems. And the folks who grow, sell and eat this kind of food are criminally unserious about talking about solutions. . . . Until we do, all this 'save the world' stuff? It's all bullshit."

Following this salvo, the thirty-six-year-old programmer-turned-farmer makes a strong push for technology. He praises robotic farming tools and vegetables grown in vertical indoor farms with very little water. He even calls for animal-free meats grown in laboratory dishes: "If technology can offer decent, affordable meat to people without environmental side-effects, I don't have the right to reject that technology out of hand just because it's new, weird, or might threaten my revenue potential."

Chris makes it clear at the end of his screed that he's not quitting permaculture farming, he just wants to broker some peace between the *de*inventors of agriculture and the *re*inventors. "So let's build our soil and grow good food," he says to his fellow plantsmen. "But let the folks in the lab do their thing, too. Like it or not, we're all depending on one another."

Chris and Annie, I would learn during my visit, are integrating into their farming new practices and tools that are anathema to some of their peers. "We're old-school as much as we're new-school," says Chris. "That's not a contradiction, it's just the best way to be." Farming's been a high-risk industry since the beginning of civilization, he observes, but in a hotter, more crowded world, "you have to farm smarter."

If there's one thing the Newmans have learned in their first half decade of farming, it's that twenty-first-century food production is all about risk—being willing to take it, and learning how to manage it. "Do whatever you can to understand the risk farmers face," Chris tells me. "What it takes to grow food under normal conditions—that's hard.

To grow food in changing conditions, that's a lot harder." With that in mind, we'll begin this story in Eau Claire, Wisconsin, at the apple farm of Andy Ferguson, a second-generation farmer who, even at age thirty-two, knows a lot about the increasing risks, and enduring rewards, of his trade.

CHAPTER 2

Killing Fields

Even if I knew that tomorrow the world would go to pieces,
I would still plant my apple tree.

—MARTIN LUTHER

ON MAY 15, 2016, Andy Ferguson woke up as he always did at 4:29 a.m., seconds before his alarm rang, and immediately noticed the frost creeping up his bedroom window. He slid out of bed, dressed quietly, poured a thermos of coffee, and lumbered to his Ford F-350, his breath producing clouds of icy mist in the predawn darkness. He wasn't too worried yet—the weather forecast had said a low of -1, and his apple trees could handle that—but the air had an ominous bite to it.

Andy drove to a digital weather station at the edge of one of his fields, which had recorded temperature ranges during the night. Around two a.m. the mercury had dipped to -3 and hadn't budged. His stomach rose. Any sustained period of temperatures under -1.5 posed a real threat. The water in the tissues of the apple blossoms could freeze, creating ice crystals that would rupture cell membranes and kill the emerging fruit.

Ferguson walked to the nearest Honeycrisp tree, pinched a blossom from its branch, and pulled a three-inch pocketknife from his work belt. With the bloom in his left palm, he slid the knife tip down its middle, bisecting the bottle-shaped belly. Inside, the apple was in the earliest stage of growth, barely a millimeter wide. "Where there should have been green living tissue, it was blackish brown," he tells me when we meet several months later. "That dark smudge is exactly what you're hoping not to see—no chance of fruit." Ferguson had seen plenty of it just a few years before, when the late-spring freeze of 2012 wiped out 90 percent of his crop. "You can't survive in this industry if you get caught up in despair," he says. "Either you roll with the punches that Mother Nature throws at you, or choose another line of work." He snapped a picture of the filleted bloom on his phone and began moving up the hill. His orchard's hilly, and he knew the trees in the lower-lying regions, where cold air collects, would be more vulnerable.

At the time, Andy and his brother and father owned three orchards in western Wisconsin, each within a thirty-minute drive of the others. Those thirty minutes—about twenty miles—were usually enough to spread their risk. If a hailstorm, for example, hit one orchard, it might spare another. The Eau Claire location is the largest, and the one where Andy built the house he shares with his wife and two daughters. It was also the site of his first kiss, his marriage proposal, and his wedding. He knows each old, craggy-limbed McIntosh, Cortland, Riverbelle, Haralson, and Honeycrisp as if by name, and every lanky whip of the new cultivars Zestar and Pazazz. In 2012, Andy had finished his law degree at the University of Wisconsin–Madison and then joined his dad in the family business. The father-son team had grown their holdings to 350 acres that during good years produced about seven million apples. They had a popular pick-your-own retail operation for locals and sold the rest of their fruit wholesale to Walmart, Sam's Club, and regional supermarkets.

Under the moonlight the previous night, the orchard surrounding Andy's house possessed an ethereal beauty. The trees had burst into a "snowball bloom," so inundated with apple blossoms that they looked covered in the aftermath of a pale pink blizzard. The winter of 2016 had been warm, the warmest on record, and the trees had begun their first stage of blooming, or "bud break," about a week early that year. Now their petals were full grown and yawning wide open. The only challenge, it seemed, was the abundance: there were so many blossoms that they'd need to be thinned by about 80 percent. Andy was particularly concerned about the whips—too many fruit would draw nutrients and energy away from their young trunks and branches, slowing their growth and reducing their yields over time. He'd been mulling this over the night of May 14 as he drifted off to sleep: forty thousand whips, eighty blooms per tree . . . some three million blooms would have to be removed within a week.

Snowball bloom in a young Ferguson orchard

The next morning, Andy was scanning trees for fruit to salvage, not to cull. Several rows up the hill, he pinched another bloom, divided it with his blade, and found the same ruptured tissue. He kept moving upward, faster now, performing the miniature surgeries on the blooms of tree after tree. He texted his dad a picture of one of his specimens and typed: "Fell to 26 [degrees Fahrenheit, or -3°C] here. Lotta bloom kill." His dad, who was at their location twenty miles south, was seeing similar damage. By midday, Andy had dissected the blossoms of more than two hundred apple trees. All were dead inside. The Fergusons wouldn't know the full impact of the freeze for several weeks—those four hours of below-freezing weather in May had killed about six million infant apples. They'd lose about three-quarters of the yield across their three farms, more than $1 million in lost harvest.

In the days after the frost, Andy spent most of his waking hours sampling and testing more blooms. He measured whatever fruit growth he could find with a micrometer, a small metal device resembling wire pincers. He compiled his data and created spreadsheets and charts to get a better feel for the damage across the three orchards. The task was Sisyphean: "I'd find maybe one in fifty or a hundred blooms with living tissue inside. I'd measure it on a Tuesday at 4.5 millimeters, then come back two days later and it might be 6.5. I'd go, Okay, that's good. Then I'd find another survivor that grew to 4.5 and then quit and died," Andy recounts. "I'd know the whole process was in vain, because if the bud's going to become an apple, it's going to become an apple, and if it's damaged, it's damaged. You can't reverse the fate at that point. You're just trying to feel some measure of control."

Soon Andy began traveling to orchards throughout the region with even deeper losses, helping growers assess their damage. A year earlier, he'd been elected president of the Wisconsin Apple Growers Association, and under this mantle he would soon petition the governor's office to declare a state of disaster and help uninsured farmers absorb the cost of their crop damage. Andy also began to investigate how to protect his apples from warm winters and late-spring frosts. The 2016 freeze was an eerie echo of the early blooming and freak frost that had devastated his yields four years earlier. If these extremes were the new normal, he and his fellow growers needed a new plan to keep apples in this region alive.

OF THE MANY wild fruit crops that have been tamed by agricultural scientists, there are few today so heavily manipulated—so far removed from their ancient roots—as apples. Modern apples bear as much resemblance to their early ancestors as, say, a drone does to the Wright Flyer. Somewhere around 1000 BC, the seeds of apple trees native to what is now southeastern Kazakhstan were dispersed throughout the world by traders along the Silk Road. But the thousands of original wild apple varieties that were eventually propagated throughout Russia and Europe, and later across the United States, have since been radically narrowed to the scant dozen or so varieties now bred for markets worldwide.

The wild trees can grow a hundred feet high and nearly as wide, and live for about a century. Domesticated trees are today deliberately dwarfed to about ten feet for ease and speed of harvest, and their productive life is no more than two or three decades. Many aren't even freestanding trees at all. Andy Fergusons's Zestar and Pazazz trees are lashed to trellises and trained to look more like vines or hedges. Wild apples range dramatically in size, color, and flavor. Some taste "spirited and racy," wrote Thoreau, others are "sour enough to set a squirrel's teeth on edge"—a far cry from the predictably sweet, mild-tasting red and green orbs most of us buy today.

The starkest difference between modern apples and their ancestors lies in the way the fruit is reproduced. Apples are heterozygous, meaning that the seeds generated by each fruit are genetically different from the fruit itself. The physical features of any given wild apple can be distinct from its parents, and that fruit's own seeds will produce progeny that represent another scrambling of the genetic code. Heterozygosity is great from an evolutionary standpoint, but not great if you want a reliable, reproducible product, which is why modern apple orchards use a cloning process to propagate the narrow range of apple varieties that are stocked in your supermarket. Branches of the parent tree are grafted on to a scion or rootstock, creating orchards filled with exact genetic replicas of each chosen variety.

To my own crunch-loving, sweet-toothed palate greedy for a year-round apple supply, there's a lot to love about the products of modern orchards. But the insects and fungi and diseases that prey on apple trees also appreciate the reliable results of cloning. Over time, an incredible range of predators, from codling moths and leaf miners to apple scab and fire blight, have developed ever smarter ways to burrow into apple trees and their fruits in all stages of growth. So far, breeders and agricultural scientists have managed to subdue virtually every one of these environmental pressures—usually with a technological fix, often in the form of chemicals. More pesticides and fungicides are applied to apple orchards than to any other fruit crop. And the brinksmanship between grower and nature extends well beyond harvesttime.

The average apple sold in the United States today has been in storage for six to twelve months before it reaches a store shelf. Apple trees produce a biennial crop, every other year, and only in autumn—but the consumer demand is constant. The development of "controlled atmosphere storage" a few decades ago enabled distributors to meet those year-round demands. The process manipulates temperature and humidity in storage facilities, along with oxygen and nitrogen levels, in turn controlling the apple's ability to "breathe" through the small pores in its skin. The storage conditions slow down the ripening process so much that the fruit can taste fresh even after idling in a warehouse for more than a year.

Remarkably, several leading studies have found little to no decline of antioxidant activity during the storage period. Another study in the *Journal of Nutrition* shows that up to 40 percent of an apple's antioxidants decline over six months, presumably under suboptimal storage conditions. Either way, modern breeding tools and long-term storage have vastly increased the fruit's availability. The "apple a day" mantra wasn't a financial or geographical reality for most American families until the 1950s. Apples are now a $4 billion industry in the United States, and more than half the global apple supply comes from China. Those perfect pyramids of glossy green and red orbs that are a mainstay of American supermarkets can be found in storefronts around the world. I've encountered apples everywhere I've traveled, from remote towns of Norway to village markets in Kenya. Everywhere it's nearly the identical product, assembled like so many sweet, colorful bricks in the same sturdy displays. The production of apples has come to seem inexhaustible, even inviolable. It's hard to imagine that these ubiquitous fruits still come from nature—and that they remain at nature's mercy. But, of course, they do.

WHEN I ARRIVE on a cloudless July day at Andy Ferguson's Eau Claire orchard two months after the devastating May freeze, it looks, to my untrained eye, perfectly healthy. The trees, planted in perfect rows, are loaded with glossy leaves; the ground between the rows is covered in thick green grass. Throughout the previous spring, I'd been researching more visually recognizable effects of climate change on agriculture—coffee rust fungus that had swept across plantations in Brazil; maize fields in Ethiopia withered by drought. The damage here in Wisconsin was visually subtle but, I'd soon learn, no less severe.

To get to Eau Claire, I'd driven from Green Bay past uninterrupted stretches of foot-high corn so wide and shiny in the summer sun they looked like bodies of water. Wisconsin is "America's Dairyland." It produces a quarter of the nation's cheese and butter and is the second-largest supplier of cow milk. A majority of the state's farmland is devoted to Jerseys and Holsteins and the cornfields that feed them. The entrance to Ferguson's farm broke the monotony of stalks with a wooden welcome sign: FERGUSON'S ORCHARD: APPLES, PUMPKINS, FUN. The fun was visible just beyond the entrance, next to a barn, where Andy had built a big wooden jungle gym and a fenced-in area for a petting zoo.

Growing up, Andy had been that rare kid who could work toward a distant reward. He started his first business, a lawn-care company with three push mowers and one employee, when he was thirteen. Lawns earned him the $4000 he needed to buy a 1984 GMC Jimmy. The truck sat in the driveway until he was old enough to get his learner's permit. After tenth grade, Ferguson turned his energy to the orchard, where he worked weekends and summers, learning all aspects of the apple trade, from grafting to bookkeeping. "I figured out early on it takes a certain kind of person to love this work. You've got to think long term and be patient and thick-skinned," he tells me. "I knew I was that kind."

Andy had graduated from law school on the dean's list and gotten job offers at two firms and a Fortune 500 company in the Midwest. They'd promised him a six-figure salary. He turned them down, following the example of his dad, who worked as a plant manager at 3M before he left corporate America to buy and tend the family's first hundred-acre orchard. Andy is almost a physical cliché for a farmer in his business: tall, broad shouldered, and apple cheeked, with hands big enough to hold six Honeycrisps in each. He looks like the apotheosis of a rising generation of midwestern farmers. At age thirty-two, he is, relatively speaking, a child in his industry, an anomaly within an aging population of American food growers.

Farm size has grown exponentially over the past half century while the number of American farmers has declined from more than 6 million in 1910 to 2 million today. The agrarian life once heralded by Thomas Jefferson as the embodiment of American ideals—"Those who labor in the earth are the chosen people of God . . . whose breasts he has made his peculiar deposit for substantial and genuine virtue"—has over time hemorrhaged participants, especially young ones. The average age of a farmer in the United States is fifty-seven, and a good portion of American farmers are well over seventy. American farm owners now make up less than 1 percent of our population, but they manage about 40 percent of our land. That land will live on long after each farmer's brief tenure tending it, and will, as much as any other factor, determine the survival and success of generations to come.

Andy, like Chris and Annie Newman, sees agriculture at a fascinating crossroads between old- and new-world strategy. "Of all industries, farmers are the closest to nature," he observes. "We think of the elements traditionally as beyond our control, but we have to be proactive. We can't fight Mother Nature—she's the source of what we grow—but we can redirect her." Andy does not associate the extreme weather that's been plaguing his business in recent years—hailstorms, early blooming, severe spring freezes, the odd summer drought—with climate change per se: "I don't pay attention to the global warming debate much, there's extremism on both sides," he tells me. Heartland farmers, in general, have not expressed much concern about climate change—as a political issue anyway. When I push Andy to address the broad international consensus on climate science, he resists: "That's beyond my frame of reference. I'm a farmer who utilizes crop science, not a scientist who farms."

He allows that "on my own farm and in my own region, at least" the conditions for farming are becoming harder to predict and less certain. He's determined to find ways to succeed within those new realities, and optimistic that he will. "Plants adapt," Andy says, "and so do we. That's probably what we as a human species do better than any other—create tools and adapt."

As he sees it, his orchards prove it: the new tangy-sweet cultivars he's growing to satisfy changing palates; the individual trees he's replaced with high-density "fruit walls" to meet higher demands; the hail netting he's begun to install over big areas of his highest-value crops—a response to a damaging hailstorm in 2015; the new generation of "pheromone traps" that seduce predatory insects with synthetic mating hormones and kill them with nontoxic solutions. Andy plans to eliminate almost all pesticides on his orchards, using this and other pest-prevention strategies. He has high hopes for robotic apple pickers that could bring down his harvesting costs. He welcomes, too, the use of genetic engineering to breed apple trees with inherent pest resistance and drought tolerance. But he says that breeding "smart" fruit trees, if it's even possible, will take time—decades, at least.

ANDY HAD INVITED me to tag along on one of his weekly scouting trips—a task he performs by truck and foot to scan his trees for signs of damage, noting and troubleshooting any potential outbreaks of pests or disease. He wears blue jeans, a T-shirt, and a baseball cap with the Ferguson's Orchards logo, and wraparound sunglasses that, on the rare occasion that he removes them, expose stark tan lines around his eyes. In stand after stand, Andy points out varying signs of distress, some related to the May freeze, others routine environmental pressures. Within a couple of hours of our scouting trip, the verdant orchards begin to look to me like the botanical equivalent of an ER triage unit, with patients suffering conditions as varied as severed limbs, lethal swelling, skin burns, and cardiac arrest.

In the Eau Claire orchard, we walk through a young Pazazz stand. The new cultivar had been developed by a private Wisconsin apple breeder as a variant of the Honeycrisp. The tissue of the Pazazz, like the Honeycrisp, has larger than normal cells that, he says, "burst open rather than cleave" when bitten, offering a juicier bite than conventional apples. The Pazazz flavor was both sweeter and tangier, with elevated sugars and acids, and Andy thinks the extra zing can make it a major commercial success. Two years earlier, he'd planted forty thousand Pazazz trees using the high-density trellis system, and already they'd begun to grow together, vinelike, to create the effect of a continuous wall of fruit.

From above, Andy's trellis walls look more like row crops than apple trees

These were the trees Andy had been concerned about thinning just before the frost. He'd ended up, of course, with the opposite problem—the Pazazz trees were mostly denuded of normal fruit. We passed one with a scattering of apples the size of navel oranges, perversely large for a young tree on a mid-July day. "Look how big this guy is already—with two months left to grow," said Andy, cupping his hand around the fruit. "By autumn he'll be massive, like a little pumpkin." The young tree, which under ordinary circumstances would have enough energy to grow about twenty full-sized apples, was driving all that energy into the few that survived the frost. The biological objective of any fruiting plant is to coax animals into eating its delicious gifts and passing on its seeds. In this case, the tree was supersizing its few fruits to better attract an herbivore, but the effort was going awry. The young tree couldn't recalibrate its energy and was "growing the fruits so fast they'll probably split right out of their skin by picking time," Andy says. For months, this image would haunt me, of a spindly tree producing huge, skin-splitting progeny in a fraught attempt to survive.

We climb in the truck and drive to a stand of eighteen-foot-tall Honeycrisps—adolescent trees seven years old and at the height of their productivity. Deep within their branches, where blooms had been insulated from the freeze, Andy points to troves of round, normal-sized apples with brown stripes around their middles that he identified as "frost rings"—scars that had been left on growing young fruit by another freeze that followed the May 15 event. Andy points out several branches that had broken from one of the trees—a sign, perhaps, of root damage. Inside other Honeycrisp trees he shows me small, misshapen apples, like little clenched fists, that also survived the frosts but with cellular damage that caused them to grow in contorted forms. He'll keep the ringed and gnarled apples on the trees, he says, to feed horses or grind into applesauce.

At one Honeycrisp tree, Andy squints at the leaves. "See how dull they are?" I couldn't—they looked fine to me. He pulls off a leaf, peers at it, and then opens a little pouch that hangs from his work belt. Inside is a microscope about six inches high from a high school biology kit

Andy had ordered online. He slides the leaf under the scope, looking in. "Mites," he mutters, showing me the evidence and noting the tree location.

Toward the end of the day, we arrive at a mature stand of McIntoshes, about twenty years old. One's a stunning sight: a thin, sickly-looking tree, shorter than the rest and mostly leafless, but the frail black branches were laden, inexplicably, with hundreds of young apples. "That's the last gasp of a dying tree—always strange to see it. . . . She knows her time is up so she's trying to pump a ton of apples out there because that's her biological purpose, to ensure the bloodline continues," Andy explains. The tree had been weakened by a viral disease, he adds, and the stress of the freeze seemed to have dealt the final blow. Because the tree is old and sick, its blooms had been delayed and protected from the frost. Here is a strange grace note—the tree's ailments allowed it to release this final Hail Mary pass of genetic information into the world before it ups and dies. Even though that information would ultimately be immaterial in this orchard of cloned trees, I feel oddly hopeful witnessing this final act—proof positive of the plant's primordial will, against increasing odds, to persist.

APPLE TREES ARE forward-thinking. They like to grow, as do cherries, peaches, and other stone fruits, in regions with four distinct seasons, and all of them need exposure to a certain amount of cold during the winter to bear fruit in the spring. These trees have learned from thousands of years of experience that the winter season is subject to occasional warm spells and other weather volatility, and they've developed cunning biological instruments to protect their blooms from damage.

During the autumn, as days grow colder and darker, the buds accumulate hormones that control a dormancy period in the winter which will inhibit growth, even during warm spells. In order to exit this period of winter dormancy, apple trees have to rack up a certain number of "chilling units," to use the term of art, each unit representing one hour of exposure to a range of temperatures just above freezing, between about 0 and 7.2 degrees Celsius. Only within this narrow temperature spectrum can a hormonal shift take place within the tree that awakens it from dormancy and gives way to new growth. The chilling period works somewhat like an insurance plan—just as your coverage kicks in once you've paid off your deductible, an apple tree can begin budding only once it's met its chilling requirements. When winter dormancy ends, the tree can begin to accumulate a certain quota of "heat units," or hours exposed to a range of warm temperatures that produce growth-promoting hormones and eventually stimulate blooms.

Apple trees generally have chilling requirements between about 800 and 1800 units, depending on their variety and region. Southern apple-growing states like Georgia and North Carolina tend to be on the low end of this spectrum, but even so, given the warmer-winter trends of late, they've been coming up short. "Underchilling" has thrown the trees' pollination process off track, causing blooms to open sporadically or not at all so that they can't pollinate synchronously and they wind up producing less fruit.

In northern apple-growing states, warmer winters have been causing, counterintuitively, a "superchilling" phenomenon. A typical winter in states like Wisconsin, Michigan, and New York exposes trees to temperatures well below freezing for long spells, and at these temperatures, the trees power down into a deep hibernation. During warmer winters, they get less of this deep-cold REM sleep and spend more time in the just-above-freezing range, where they accumulate extra chilling units. Trees that have been superchilled become more vulnerable to blooming. Like a kid who's been overserved candy and becomes increasingly prone to tantrums, a superchilled tree gets too much of a good thing and becomes overstimulated, needing fewer than normal heat units to begin spring growth.

That's probably what happened to Andy Ferguson's apple orchards in the spring of 2016. The trees had amassed extra chilling units during the tepid winter and when early April brought on a warm spell, even a brief one, they were overeager to bloom. The trees misread the temperature signals as an indicator that spring had come. The blooms began to push their petals through the protective bud casing in the first days of April, about a week earlier than usual, making them vulnerable when the freeze hit.

The orchards of the Upper Midwest were not the only ones that got confused by the strange weather in 2016. Record-breaking warmth and temperature swings had swept New England months earlier, causing on February 14 what horticulturalists dubbed the Valentine's Day Peach Massacre. Peach trees in New Hampshire, Connecticut, and Rhode Island had begun blooming not six days ahead of schedule but nearly five weeks early. By February 13, many of the orchards were in full bloom, and on Valentine's Day, temperatures plummeted to 25 below zero. The three states experienced a "total kill"—100 percent loss—of that year's peach crop. New Jersey, the fourth-largest peach-producing state, lost about 40 percent of its peaches and New York's Hudson River Valley lost 90 percent. The cost throughout the region totaled about $220 million in lost peach harvest.

"You look at the fruit crop damage in recent years and think, this is happening on a scale and at a frequency that is fundamentally new," Amaya Atucha, a fruit horticulturalist at the University of Wisconsin–Madison, tells me. "We've been living in a period of climate stability for millennia and we're witnessing a shift. Conditions are changing, not just for apples but for grapes, cherries, peaches, cranberries, blueberries. And not just in a few regions, it's everywhere—disruptions are everywhere." Jerry Hatfield of the U.S. Department of Agriculture (USDA) describes the phenomenon in starker terms: "It's not just one total-kill event in the fruit industry every seven or ten years, it now happens regularly—we simply have more horticultural blood on our hands."

Fruit trees are thrown by not just warmer winters and earlier blooming periods, but also temperature swings. If temperatures careen from the low-20s degrees Celsius to just below freezing in the span of a few days (as they've done on several occasions during recent winters in Tennessee), it puts considerable stress on flowering trees. Over time, fruit trees have developed a process of hardening or "acclimation" to help them survive severe weather. Their tissues gradually build endurance against the cold much the same way that muscles build strength, layer by layer. As cold sets in and daylight shortens, the tree's bark and buds develop chemical components that prevent the formation of ice and protect them from cold. Just as a bodybuilder can work up to 400-pound weights with practice, apple trees can develop resistance to well-below-zero temperatures by means of this gradual chemical barrier building. But an abrupt transition from 100 to 400 pounds would injure any weight lifter, and fruit trees, likewise, can't easily shift from mild weather to bitter cold. Even during the winter dormancy phase, sharp temperature swings can damage hibernating buds, making the blooms sickly and fragile when they emerge.

The roots can also suffer. During normal winters, snow in most apple-growing regions acts like a blanket that seals in soil warmth and protects roots from cold damage. But frequent temperature swings can melt away whatever snow manages to fall, leaving the tree roots more vulnerable when severe cold blows in. Damage to roots is ultimately much more concerning than damage to buds, says Andy, killing "not just a single harvest, but an entire tree that might otherwise have twenty more years of apples to give."

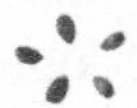

THE YEAR WISCONSIN lost its apples and New England lost its peaches also brought many other unprecedented climate-related events in the United States and throughout the world—deadly floods swept Texas; the Antarctic ice shelf cracked; portions of India reached a record-breaking 53 degrees Celsius; forest fires burned swaths of Alaska; twenty-eight inches of snow inundated New York City in two days. It was also the year that Donald Trump was elected president with a platform outright rejecting climate science. A map of the states that elected Trump to office shows every major agricultural state except California in his favor, including the swing state of Wisconsin. Gil Gullickson, a reporter for *Successful Farming*, an agribusiness trade magazine, who has covered climate-change impacts in "Big Ag" states, tells me that while "most farmers recognize that the weather trends are out of whack, many still accept the prevailing conservative attitude that climate science is bunk."

The agricultural research that is coming out of these states, however, tells a very different story. Every major ag company, from Monsanto and Syngenta to Cargill and John Deere, had accepted climate science well before 2016, and they've been building research divisions and product portfolios specifically addressing its impacts. Some of the country's leading agricultural universities were meanwhile producing studies that showed the effects of climate variability on food production in their regions, sometimes with striking results.

One such study, on cherry production, came from a team at Michigan State University (MSU). Michigan is the third-largest apple-producing state in the country and is first for cherries. Apple and cherry trees have similar blooming behavior, though the latter tend to have lower chilling requirements and thinner bark, which makes them even more sensitive to temperature changes. During the winter of 2012, the state's cherry crop had gotten walloped. After a warm winter with a straight week of 20-degree weather, trees broke bud in early March, but this was followed by an April freeze that killed the blossoms on virtually every single cherry in the state. "It was a half-billion-dollar loss for Michigan's fruit industry, and cherries were especially bad off," recalls Jeff Andresen, an agricultural meteorologist at MSU. Apple yields were down by almost 90 percent.

Frozen apple blossoms

But what was a trauma for fruit growers was a boon for

research. Before the freeze hit, Andresen and a team of grad students had spent a year collecting weather data to examine shifts in blooming periods of cherry trees over more than a century, from 1895 to 2013. They'd culled the data from weather-sensing equipment throughout the state and from recorded government archives. After the 2012 freeze, Andresen's study drew additional funding and interest. The team redoubled their efforts to parse the data in three categories: blooming periods for cherry trees, spring freeze events, and rainfall levels. The results, which they first released in 2016, were stark: in all categories they found stable data during the fifty years from 1895 to 1945, but in the decades after World War II, trends noticeably changed.

The blooming data jumped off the page. "There are a few times in your career when data stands up and asks to be noticed. This was one of them," Andresen tells me. His team found that the early budding process had shifted more than ten days earlier in spring over the seventy-five years following World War II, a period of time during which there had also been steep increases in greenhouse gas emissions. In 1945, the normal average date for the "green bud stage," just before the petals push out, was April 5. By 2013, the date had crept to March 26.

The frost data also surprised Andresen. Before 1940 there had been no more than ten spring freeze events per year. In the decades after 1940, that number reached nearly twenty spring freeze events, five more per year on average. "They weren't all severe or damaging freeze events, but they were more frequent"—a sign of greater volatility, he says. The last category, rainfall levels, was critical to the cherry industry because the thin skin of the cherries, if exposed to even light rain just before harvest, will crack. Andresen and his team found on average a 10 percent increase in the volume of spring rainfall since 1945.

"I'd heard plenty of growers tell me that springs were wetter, that crops were blooming earlier, and freezes were more frequent," says Andresen, "but to see the numbers tell that story—to see the cherries tell a story about broader temporal trends—it rarely happens in my field." When I asked Andresen about the larger implications of his data, he responds, "The upshot is, if you're a grower who wants to succeed, it will require major intestinal fortitude."

This is especially true if you're a peach farmer in the American Southeast. Dr. José Chaparro, a professor of horticulture sciences at the University of Florida, says that warmer temperatures are bringing new pests and pathogens to the farms in this region, while warmer winters and inadequate chilling hours are causing smaller fruit size and poor shape. As with apples, underchilling causes reduced and sporadic blooming, which he says is "a production nightmare." Chaparro warns that the peach industry in Florida, Georgia, and other southeastern states could collapse if warming trends intensify, and he's racing to breed new varietals that can withstand the heat. His team is selecting for traits from "low-chill" peach trees that need fewer chilling hours than most of those that are commercially grown. "We are essentially hoping to redesign the peach tree," says Chaparro, "to survive the new normal."

FRUIT GROWERS ARE not the only American farmers struggling with climatic shifts. Farther west, Jerry Hatfield of the USDA has spent most of the last decade examining similar temporal trends related to corn farming. Hatfield dug up decades of data about spring precipitation in central Iowa over two time frames. He found that between 1900 and 1960, there were only two years when more than 1.25 inches of rain fell in a single day. From 1960 to 2017, seven years had eight or more days hitting that high-rainfall threshold. Springs were getting wetter overall, increasing the challenges of weeds and pests, as well as the spread of fungus and crop diseases. The rainfall, crucially, was also arriving in more intense downpours. Hatfield found that the number of workable field days in April through mid-May decreased 3.5 days between 1980 and 2010—the soil on cloudburst days was too sodden to get anything done with heavy farm equipment. "Even a single day of missed fieldwork can translate into significant losses," says Hatfield.

The western and southern regions of the United States have been facing altogether different challenges than those of the Midwest and Northeast—not freak freezes or wetter springs, but heat waves, forest fires, and punishing droughts. Past a certain temperature, plants have to cool themselves by respiring more water, which robs energy from photosynthesis and cuts yields. If there's not enough water to support the plant's cooling process, it suffers heat scald and dies. "For most farmers in the United States, the major limiting environmental factor in their crop production is increasingly becoming water availability," says Hatfield. The 2011 Texas drought caused a $5 billion loss in production of livestock, cotton, corn, wheat, and peanuts. The years 2012 through 2014 brought the "New Dust Bowl" to the southern Great Plains; heat waves and drought killed off tens of thousands of acres of wheat and corn. Crop insurance payouts totaled about $30 billion during those three fraught years, with U.S. taxpayers footing most of the bill. Then 2015 brought the brunt of the California drought, which cost the state billions in lost agricultural production, and nearly twenty thousand jobs.

In the aftermath of these droughts, the Departments of the Interior and Agriculture released climate reports that predicted average temperatures in the American West would rise by 3 to 4 degrees Celsius by the end of the century (more than four times the change in average temperatures we've experienced so far), causing megadroughts in this region beyond anything we've yet experienced. The reports also predicted a sharp decline in the snowpack that feeds western reservoirs, and up to a 20 percent decrease in the stream flow of the major western river basins, including the Colorado, which supplies water to the thirsty southern half of California and six other states.

"We're really just beginning to grasp the full magnitude of the problem," says Margaret Walsh, a senior ecologist in the Climate Change Program office at the U.S. Department of Agriculture. "There's a huge range of variables to consider across different regions—temperature shifts, precipitation changes, invasive pests, new diseases, sea level rises." Walsh emphasizes that some of the worst impacts may come from disturbances in global supply chains and distribution networks.

It's too early, she says, to start prescribing solutions to a problem we don't yet fully understand. And anyway, it won't be primarily a top-down effort. The most effective solutions will emerge locally, region by region, built from the ground up by institutions like MSU and Iowa State University in Ames and by farmers like Andy Ferguson, engaging in trial and error, each player adding another link in that long chain of experiments stretching over ten millennia that's aimed at securing and expanding our food supply.

CENTRAL CALIFORNIA IS covered by a seven-million-acre checkerboard of farms growing more than half of all the fruits, nuts, and vegetables produced in the United States. It seemed, until recently, like a region so rich in fertile soil, so mild in temperature, and so high in productivity as to be a kind of modern-day Fertile Crescent. Almost all—between 80 and 95 percent—of the strawberries, almonds, and grapes grown in the United States come from central and northern California. That's also true for walnuts, pistachios, figs, lemons, broccoli, rice, artichokes, potatoes, and tomatoes, as well as for three-quarters of our leafy greens. The dairy cows that graze in California pastures produce 20 percent of the national milk supply. *The New Yorker*'s Dana Goodyear has described the midsection of the Golden State as the "fruit basket, the salad bowl, and the dairy case of America."

But by 2015, three years into a withering drought and after six straight months with no rainfall, much of central California's most fertile farmland was bare and bone dry. Reservoirs in the region had sunk to half capacity and utilities were rationing irrigation water. More than half a million acres of crops had been fallowed. News coverage showed images of exposed lake beds strewn with old sofas and rusted cars, of desiccated fruit and nut orchards in which the naked trees looked like outstretched hands begging for rain. The media told stories of strange and unexpected losses: California beekeepers had to resort to feeding their bees sugar syrup and processed bee food (a blend of oil and pollen) because nectar-giving crops were so compromised. The prized wines of Northern California were also taking a hit. Wine grapes like heat, but too much of it can cause a kind of thermal shock that throws off their flavor.

Perhaps for the first time, the impacts of drought and climate change on food production rattled mainstream America. The subtle changes that had been occurring for years on heartland farms growing corn and soy had been easy to ignore—now, suddenly, strawberries and Chardonnay were on the line.

A few years later, in 2019, summer heatwaves pummeled France's wine industry, cutting production by more than 10 percent. The same year, Australia's wildfires scorched livestock production, killing millions of sheep and half a million cattle. By then, specialty food producers worldwide were feeling the heat. Extreme weather had taken a toll on the avocado farms of Mexico and the pistachio farms of Iran; bacterial blights had destroyed millions of acres of olive groves in Italy. Even chocolate had come under threat—warming trends were aggravating problems such as "frosty pod rot" and "swollen-shoot virus" that are caused by tropical plant pathogens on the cacao farms of western Africa and Central and South America.

And while many of us may be able to adjust to life without pistachios and avocados, maybe even willing to cut back on chocolate and wine, one key import is almost universally indispensable: joe, java, rocket fuel—coffee. I shudder to think what would happen to economic productivity in my own household, let alone to the national GDP, if the dark roast ran out. But in 2016, the Climate Institute released a report detailing how warming trends are bearing down on the coffee farms of the key countries in the "bean belt," including Ethiopia, Brazil, Colombia, Mexico, and El Salvador. Coffee plants thrive in the dry, relatively cool conditions that can be found in the highland farming regions of these countries, but rising temperatures have begun to slow and stunt berry growth. Hotter weather is also fueling the introduction of pests such as the coffee berry borer and of a virulent fungus known as leaf rust. Leaf rust has already killed off millions of acres of coffee crops in South America alone. The Climate Institute report predicts that, given current warming trends, by 2050, the amount of suitable farmland for coffee will be cut in half—unless we find ways to adapt.

International coffee companies have helped form and fund World Coffee Research, an organization of scientists developing a range of responses, including shade management and mulching, which can cool the plant's roots. The group is creating a gene bank to preserve the genetic diversity of wild coffee plants, which over many millennia have learned to adapt to stressful climatic conditions. Benoît Bertrand, a lead researcher for the organization, is hopeful that genetic information contained in the seeds of Arabica's wild ancestors can be harnessed with modern tools and bred into new plants able to adapt to the pressures ahead.

IN ROBERT FROST'S 1920 poem "Good-bye, and Keep Cold," he begs his apple orchard not to succumb to late-winter warm spells while he's off tending the maples. It's a reminder that spring frosts have been freezing orchards as long as there have been orchards to freeze. The difference today is not the presence of the threat, but its scale and increasing frequency, and the urgent search for solutions.

Around the time of Julius Caesar, some two thousand years ago, during the early golden era of viticulture, the winemakers of the Roman republic got the idea of heating their vines on cold spring nights to protect the grapes from frost damage. In between their trellises they built small fires made of prunings, dead plants, and other farm waste. Eventually, orchard keepers throughout Europe and the United States caught on to the practice, building open fires among their rows of apple, peach, and plum trees when spring nights threatened to freeze. There's no evidence that these efforts were effective, but we do know that the practice persisted. By the twentieth century, growers in the United States heated their crops by burning sawdust mixed with heavy oils or old rubber tires in metal casks, devices that came to be known as frost candles. The fires produced an unctuous smoke that the growers hoped would coat the buds with an oily film and help block their heat loss. It didn't work, and the smoke was so polluting that by 1970, the practice was banned in the United States.

Outdoor crop heaters today are a niche market. Andy Ferguson recalls seeing his neighbors during the freeze of 2012 hauling around Frost Dragons—propane-fueled heaters attached to huge fans that are pulled through orchards by tractors. The heat-blowers can raise temperature in the orchards a degree or so if they're dragged every eight minutes through each row, according to the Frost Dragon website. But while the crop-heater business has grown in recent years, in particular among vineyard keepers and in pistachio orchards to protect from mild Californian frosts, there's still little scientific evidence that the devices work.

Andy, for one, sees this technology as an expensive crapshoot, no more reliable than "trying to heat a house with a candle." But in his research following the May 15, 2016, freeze, he turned up some more promising options. Michigan cherry farmers had been having success with hydrocooling, a process of applying a fine mist of water to buds during warm winter spells; as the mist evaporates from the buds, it produces a cooling effect and prevents them from losing their chilling units. (As part of his cherry study, Jeff Andresen researched this method and found that it could delay cherry budding by up to a week or more. "If it had been used during the winter of 2012, I'm pretty certain hydrocooling could have averted most of those crop losses," he tells me.) But installing the overhead irrigation infrastructure is expensive and time-consuming, and the method hasn't yet been proven to work on many large commercial apple farms.

Andy settled on another solution that seemed to him the most viable: 40-foot-tall, 130-horsepower wind machines called frost fans. Warm air rises when temperatures drop at night, creating a tier of warmth known to meteorologists as the inversion layer just above the colder air at ground level. Frost fans are designed to pull air down from the inversion layer, blending it with the cold. Andy had read a Cornell University study that found that the machines could reliably increase air temperatures by one to two degrees within a particular range of cropland, each covering a span of about ten to fifteen acres. "It won't work if you have temperatures that go way low," says Andy, "but on a night like May 15 when you're getting down to 26 [degrees Fahrenheit, or -3°C] and want to keep your temps at 30, it can make all the difference." Andy found that between 2014 and 2016 the number of frost fans on orchards in Michigan and New York had jumped by a factor of ten—from about fifty machines in each state to about five hundred. He was sold.

Andy Ferguson

In August 2016, the Fergusons purchased three frost fans for $35,000 apiece. They arrived at the Eau Claire orchard on September 17. In the days leading up to delivery, Andy drove his mini-excavator into the Pazazz stand and dug up thirty trees in three different locations. He scooped out three holes, eight feet square and three feet deep, and poured thirty-three thousand pounds of concrete per hole. A technician from Orchard Rite, the manufacturer, brought in crane equipment to erect, wire, and bolt down the hulking fans. The installation was bittersweet, Andy says. "It felt good to be adding control measures, but I'd much rather that investment be in new crop volume. I'm used to spending this kind of money to add value to our farm, not to avoid losses." Andy also pursued some low-cost measures, like planting winter cover crops to insulate the soil in the absence of snow, and applying special paint to the trunks of trees to reduce bark splitting during freeze events.

I was by this point in my research impatient to find strategies that would protect the fate of Andy's apples, but frost fans smacked of, well, tilting at windmills. The notion of blowing fans in the open air seemed no more tenable or energy-efficient than using frost candles to heat the outdoors. I cringed at the thought of farmers burning electricity to blend air layers when the planet's whole freaking atmosphere is out of whack. But from Andy's perspective, that's myopic. The frost fans aren't meant to be a comprehensive solution, but a bridge technology to more nuanced and sustainable solutions down the line. And there were plenty of other approaches in the meantime that were far more quixotic. Some big apple orchards in Michigan, for example, had rented fleets of helicopters to fly over their trees and push down the inversion layer during freeze events. The choppers could cover up to forty acres apiece, but the cost was astronomical, with an hourly rental fee of about $1600 per craft. "When you're looking at potentially huge losses in your crop value overnight," says Andy, "you're willing to resort to expensive measures."

AFTER MY VISIT, Andy and I stay in touch. We swap emails with links to articles about the climate-change impacts on fruit production, and on new technology that's emerging in the field—some dubious, others more promising.

I read about some wacky-sounding stuff: the development of "cryoprotectants," for example, which are essentially antifreeze compounds found in organisms including Antarctic insects, fish, and amphibians that enable them to endure extremely low temperatures—chemicals that might also help protect the delicate tissue of an apple blossom during a freeze, but hadn't yet proven effective. About researchers at Washington State University who had successfully tested "nanocrystals" that can coat and protect fruit buds in spring to prevent frost damage, and others at Michigan State formulating "growth regulators" that mimic the hormones produced within the bud, and when sprayed on fruit trees can prolong the chilling periods and delay flowering. Questions loom, though, about the safety, efficacy, and cost of these compounds.

"Here come the robotic apple pickers," I write to Andy at one point with a link to a company named Abundant Robotics. They've developed a robot with multiple arms that look like vacuum hoses. The robot "sees" the apples via cameras, determines which are ripe, swoops toward the apple, and gently sucks, grips, and plucks it off the tree. I call up the company's CEO, Dan Steere, who tells me the robot will be on the market "in the next few years" and that "robotic harvesting for pretty much everything you see in the produce section could be a reality inside a decade." Andy is happy to hear this because he's been facing a labor shortage. "We pay our top pickers twenty-five dollars an hour but it's backbreaking work, and not many people want to do it," he says. "My two biggest problems are weather and labor. Maybe this'll solve one, so I can focus on the other."

In the summer of 2018 Andy and his brother and dad purchased a large apple farm in Minnesota, doubling their holdings. It's part of his strategy, he says, for "continued separation of our orchards for protection from extreme weather events." The thirty-two-year-old farmer now has more than a quarter million trees on 300 acres in two states.

"You keep asking for what motivates my optimism and I feel like I never really have a good answer," he tells me during a follow-up call. "You probably wanted some profound quote or something, but if you really wanna know what gets me going, it's this—I'm texting it to you." Up pops a link to a Dodge Ram commercial that had aired years earlier, during the 2014 Super Bowl. The ad featured "So God Made a Farmer," a 1978 speech by the conservative radio broadcaster Paul Harvey. I put Andy on speaker and stream the ad, which has more than twenty-three million views on YouTube. The speech crackles with the sound of vintage radio:

> "And on the eighth day, God looked down on his planned paradise and said, 'I need a caretaker.' So God made a farmer. I need

somebody willing to sit up all night with a newborn colt. And watch it die. Then dry his eyes and say, 'Maybe next year.' I need somebody who can shape an ax handle from a persimmon sprout, shoe a horse with a hunk of car tire, who can make harness out of haywire, feed sacks and shoe scraps . . . who will stop his mower for an hour to splint the broken leg of a meadowlark. It had to be somebody who'd plough deep and straight and not cut corners. Somebody to seed, weed, feed, breed and rake and disc and plough and plant and tie the fleece and strain the milk . . . who'd bale a family together with the soft strong bonds of sharing, who would laugh and then sigh, and then reply, with smiling eyes, when his son says he wants to spend his life 'doing what Dad does.' So God made a farmer."

CHAPTER 3

Seeds of Drought

Truth is born into this world only with pangs and tribulations, and every fresh truth is received unwillingly. To expect the world to receive a new truth, or even an old truth, without challenging it, is to look for one of those miracles which do not occur.

—ALFRED RUSSEL WALLACE

PAUL HARVEY, we can safely assume, did not have a seventy-two-year-old Kenyan woman in mind when he imagined God's consummate farmer, but Ruth Oniang'o fits his description and, in important ways, surpasses it. Oniang'o has deeper reserves of gentleness and tenacity than any farmer I've met, and she lives a long way from the American heartland—about fourteen thousand miles.

Ruth grew up in Emuleche, a small village in western Kenya. Her family's two-acre farm sat on a hillside where the soil was poor and gravelly. By the time she could walk, Ruth was helping her mother grow maize, finger millet, sweet potato, sorghum, bush beans, cowpeas, peanuts, Bambara nuts, and bananas. They had their hands and a pickax-like tool called a *jembe* to work with—no tractors or ploughs. Ruth and other kids from her village spent most of their free time gathering wild fruits. The forests surrounding the village teemed with indigenous black plums and guava. They also collected wild leafy greens that grew at the edge of the nearby river. Ruth remembers them as "swamp vegetables—they had no botanical names. It was just the food of the forest, and we had all we could ever want."

When Ruth was ten, famine swept western Kenya, a rain-rich agricultural region known for growing most of the country's maize supply. Maize (aka corn in the United States) is grown here not primarily for cattle feed or corn syrup, but for direct human consumption. It provides more than half of the calories consumed nationwide. The kernels are milled and boiled into a thick, claylike porridge called *ugali* that's eaten at most meals. Throughout Ruth's childhood, droughts came to the region rarely but reliably, every ten to fourteen years, and were ruthless when they struck. The drought of 1955 wiped out almost all of Kenya's maize crop. The famine became known throughout the country as the Hunger of the Cup because the harvests were so scarce that maize kernels had to be rationed among families by the cupful rather than by sacks or bushels. Ruth's family was among the few able to supplement their rations with the salary her father made as a policeman. She still remembers watching her cousin getting the telltale signs of kwashiokor, a disease from malnutrition—distended belly, hair loss, and scaly skin. She recalls the two elderly men in the village who'd lost their families coming to their door at every meal and holding out their plates. "My mother would say, 'Our gate is always open.' We never let a neighbor starve. I understood early in my life that hunger steals dignity."

While the Barwa family (Oniang'o is her married name) beat back famine, there were hardships they couldn't stave off: five of Ruth's ten siblings died in toddlerhood of malaria. She got the disease, too, but just after the German pharmaceutical company, Bayer, had released the drug chloroquine, which was distributed to the lucky few by relief workers. When Ruth recovered, she promised her mother three things: "I would grow up to save lives. I would build a hospital for Emuleche. I would have twenty children to make up for the ones she lost." In her adult life, Ruth would have five children, not twenty—"I'm still hoping to have more," the grandmother of seven jokes—but she kept her other two promises, and then some.

Oniang'o rose to the top of her class in high school and got a scholarship to study in the United States as an undergrad at Washington State University. She stayed on to get a master's in biochemistry and nutrition, returning to Kenya in the seventies to become a lecturer in nutrition and public health at Kenyatta University in Nairobi. She was soon tapped by the Kenyan government to help develop national food security policies. She joined the Parliament, where she served for five years. She became an adviser to the United Nations and editor in chief of the *African Journal of Food, Agriculture, Nutrition and Development*. Along the way, she founded ROP—Rural Outreach Program of Africa—a grassroots effort that today works with thousands of small farmers in western Kenya to improve their yields and livelihoods.

ROP has twin goals: increasing agricultural productivity in Africa while also protecting the interests of small-scale farmers. These goals would seem to be contradictory, especially considering the partners Oniang'o is working with to achieve them. They include the African Agricultural Technology Foundation, a nonprofit group that collaborates

Ruth Oniang'o

with Monsanto, which even after its 2018 acquisition by Bayer (the same company that produced the lifesaving malaria remedy) is widely regarded as enemy number one of small farmers. It's also funded by the Bill and Melinda Gates Foundation, a philanthropic group that has been criticized for pushing Western technology on vulnerable populations. "I am not advocating for the industrialization of yesteryear—the kind of medieval, heavily polluting agriculture practiced over the last century in the United States," Ruth tells me in our early conversations. "I am talking about using technology—modern seeds, modern methods—to benefit humanity, to produce food that's clean, abundant, and climate-smart, in a way that frees small-scale farmers from drudgery. We shall industrialize our food production while maintaining the core of who we are."

When I get an invitation to meet Ruth and some of the farmers she works with through ROP, I accept, although I'm wary. It seems doubtful to me that the tools of Western agribusiness, Monsanto's seeds in particular, could serve the interests of these rural communities. But I would come to realize during my travels how narrow the American perspective on industrialized agriculture has become. I'd learn that outside the United States, and especially in emerging economies, the debate around technology and agriculture—including GMOs—is not about better labeling for corn chips, or even about corporate control of the food system, it's about progress and, ultimately, survival.

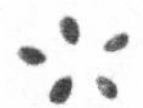

IT'S A MILD, breezy July morning in western Kenya, during the region's arid winter season, and I'm bumping along a dirt road a few miles south of the equator in an old Nissan Pathfinder with busted shocks. I'm traveling with ROP staffers from Nairobi and we've gotten lost on our way to the village of Navakholo. There are no street signs in these parts, but nearly everyone has a cell phone, and Kenyatta, our cowboy-hatted driver, is on his, gathering intel. "*Did you say, look for the tilting banana tree on the left?*" he shouts in Kiswahili into his flip phone above the roar of the engine, trying to find the correct turnoff from the main road. We turn onto a rutted path only inches wider than the truck and Kenyatta says to himself, "*Ndiyo,*" yes. This new track seems promising.

I'm slouched in the backseat next to an open window, taking in the smells of burning maize husks, ploughed earth, animal dung, diesel fumes, and eucalyptus trees. The combination is oddly fortifying. Three days have passed since my last meal, a bowl of goat stew that threw my first-world digestive system into a violent revolt. My appetite is gone, but so is my restlessness. I've packed away my iPhone and stopped trying to photograph every regal-looking woman transporting a basket of bananas or a jerrican of water on the crown of her head; the young kids moving mountains of kindling on ox-drawn carts; the monkey-filled acacia trees and baboons crawling along the forest edge; the mud hut with the sign PLANET COMPUTER CENTER; and the lone, windowless shed painted with the words PALACE HOTEL.

This region we're driving through is, you might say, the Iowa of Kenya, but of course it looks nothing like America's corn belt. Each maize farm we pass is about an acre in size, some with an adjoining vegetable garden and a scattering of animals. The fields are bordered by fences made from plantings of columnar cacti. The maize on most of these plots is ailing, the yellow-brown leaves curling from drought, chewed by pests, pocked with diseases, or strangled by *striga*, a rampant parasitic weed. Here and there we pass a farm with stalks that are suddenly, almost luridly green.

We arrive eventually at the farm we came for, the homestead of Michael and Amani Shiyuka. Their farm is bigger than most, more than two acres. Beyond the courtyard, where half a dozen people in matching white T-shirts are preparing for today's event, there's an expanse of emerald green maize with tawny gold tassels that look ten feet high. The crop, I've been told, has consistently produced annual yields about a third more productive than most of the neighboring farms, even during the recent drought, which has made the Shiyukas into local celebrities. They've used the profits from their maize sales to build a two-room house, the biggest in the area and the only one made of brick, which their neighbors call the Tego Palace. Michael smiles at the praise while

Amani, a quiet woman wearing a small cross pendant, is visibly humbled by it.

Three years earlier, the Shiyukas had joined ROP in partnership with Water Efficient Maize for Africa (WEMA), a program that had given them and thousands of other smallholder farmers throughout Kenya and in neighboring countries access to high-tech maize seeds. The DroughtTego seeds they grow (Tego is derived from the Latin word for "shield") were engineered by Monsanto. After acquiring Monsanto, Bayer has remained a major partner in the WEMA alliance. In 2018, the Bill and Melinda Gates Foundation agreed to invest $27 million in the next phase of WEMA research and outreach through 2023. Other WEMA investors include the Howard G. Buffett Foundation and the US Agency for International Development (USAID). The Nairobi-based African Agricultural Technology Foundation oversees WEMA, producing and testing Monsanto's seeds and supplying them to local distributors. The Shiyukas had agreed to let ROP and WEMA host a "field day" on their farm to demonstrate the success of these seeds and educate local farmers about modern farming practices.

"Here you are!" shouts Ruth when we roll up in the Pathfinder. She's wearing a bright polka-dot skirt and a fuchsia *kitambaa,* or head scarf, and purple rubber wellies. The banner she's just hung at the courtyard entrance reads "ROP: Eradicating Poverty Through Agriculture." "*Jamba,* Dr. Oniang'o!" I shout back, vaguely aware that I've mispronounced the greeting, but unsure why I am suddenly the laughingstock of a group of kids nearby. (While *jambo* means "hello and welcome," I later discover that *jamba* means "to fart.") So begins my long acquaintance with the venerable scholar and activist.

As Ruth and Michael lead us around the property, musicians are filling the courtyard with festive *kamba* beats. Some two hundred farmers stream in for the opening ceremony and settle into foldout chairs under big white tents. The chairs have been set up around a central stage, where local leaders and groups will soon give speeches and share performances about the challenges and successes of the farming economy. A group of thirteen girls from a nearby school where ROP staffers have been teaching farming skills begins the proceedings by marching around the perimeter of the crowd in military style, pumping their knees and arms with each step. They form two lines in front of Dr. Oniang'o, who's seated in a chair of honor in the first row of the audience.

"We stand before you, pupils of the Blessed Academy, to recite our poem 'Mkulima Bora si Bora Mkulima'—'Not Just a Farmer, but a Good Farmer,'" announces the oldest girl in the group, speaking in both English and Kiswahili. The girls range in age from six to fourteen and wear their school uniforms, burgundy jumpers with collared blue blouses. All have braided hair and carry themselves with the poise of Olympic gymnasts.

"*Wali Mkulima! Mbona munateseka, munateseka!*" they begin in unison; an event staffer whispers the English translation as they continue:

"Oh farmers! You have suffered, you have suffered!
Many farms. Poor harvests,
Poverty is all over the country.
Citizens are crying, why are you suffering?
Teachers are crying, police are crying, doctors are crying.
We are posing this question to farmers:
Why is it that there is farmland,
but yet we still suffer from hunger?"

Audience members voice their approval—"*Naelewa!*" I understand, shouts one; "*Kweli,*" true, chants another—as the girls continue.

"Poverty is not the problem, the problem is with the farmer.
Now Dad and Mum use good maize seed,
for the children to be salvaged, because of ROP, ROP, ROP Africa.
Let the farmers rejoice, because you have found a good maize.
Poverty has been defeated, for you will gain good yield."

Bananas for Ruth

The crowd applauds and the girls gather around Dr. Oniang'o. The smallest one places a gold-and-red tinsel garland around Oniang'o's neck and the tallest offers her a big cluster of bananas. Ruth stands, thanking and hugging the girls. "All my hope is in the children," she tells me later. "School intervention is the most effective work we do. It's the young ones who will drive innovation in agriculture." They are also the ones who will inherit the worst impacts of climate change, which is, adds Ruth, "our newest challenge."

THE RIVER THAT runs through Emuleche, where Oniang'o picked wild vegetables as a child, has completely dried up—one of several symptoms Ruth lists among the recent climate impacts she's witnessed. Since 2000, the Horn of Africa has experienced the worst droughts on record. At the time of my visit to Navakholo, about twelve million people in Ethiopia, Kenya, and Somalia are at risk of hunger due to recurring droughts. Ten million were receiving emergency food aid, and both Somalia and Ethiopia were facing the threat of widespread famine for the second time in six years. Kenya had suffered less from drought than its neighbors, but in 2016, the year of my visit, its maize production was down by about 10 percent due to drought and pests. We didn't know then that the following year the drought would intensify, and in May 2018, the United Nations would declare 2.4 million Kenyans severely food-insecure.

Kenya's conditions mirror what much of the African continent is up against. The country's average temperatures over the last fifteen years have been consistent with global trends—the hottest on record. Pest populations have risen and crop diseases have spread. Droughts have become longer and more frequent while rainfall has come in shorter and more intense bursts, and in less predictable patterns. The glaciers on Mount Kenya, the country's highest mountain, have been shrinking. Fewer than half of the sixteen glaciers that existed a century ago on the mountain remain intact today, and all are expected to disappear in the next thirty years. The water shortages have been particularly hard on the country's livestock population, which has declined sharply since 2000. Farmers have been trading out cattle for camels, which can better tolerate heat and drought. There are now around three million camels in Kenya, three times more than there were a decade ago.

Climate change is also driving an interest among rural farmers in new knowledge and tools. "More farmers are showing up at our field days looking for information. When their growing conditions become difficult they must seek better practices," Oniang'o tells me. During the field day event at the Shiyuka farm, the visiting farmers rotate in groups of fifteen to twenty through workshops taught by ROP and WEMA staffers. The workshops span old and new farming methods—compost science, crop rotation, and care of indigenous vegetables alongside introductions to high-tech storage materials and DroughtTego seeds. Ruth and I join a group of seventeen farmers, twelve of them women, in the DroughtTego workshop. "Most of you have experienced the shifting seasons," a WEMA field instructor tells the group in Kiswahili. "The rains fail to come—you wait and wait. Then too much comes and washes out everything. You don't know when to plant, when to weed, when your maize will be ready." There are nods and murmurs of acknowledgment. He shifts to English: "The white man calls it climate change." He asks the group to repeat the English term together. "Cliiiimate chaaange," they chant in unison slowly, and the words come out sounding like the Latin name of an obscure species. "Come the droughts, come the heavy rains, these crops can withstand climatic changes," the instructor continues, explaining that the maize has been developed to better tolerate stress and to mature rapidly—in about three months, as opposed to the typical four—and also drying quickly at harvesttime, an advantage when rains are hard to predict.

Oniang'o chimes in. "It is an odd concept at first—climate change," she tells the group. "But you have been living with its impacts for many seasons. Now you see what can happen when you try new seeds. Seeing is believing." The real purpose of the field days, Ruth tells me, is to help farmers get comfortable with new concepts and tools: "They want to plant what their parents planted, because it's what they know."

The WEMA event is designed to encourage rural subsistence farmers to modernize, a goal that has significant implications not just for the farmers themselves, but also, inevitably, for Monsanto and its parent company, Bayer. "The potential to improve yields in Africa is huge. Most farmers in Kenya and throughout the continent are using seed types that haven't been grown in the United States since 1920, 1930," Robb Fraley tells me several weeks before I travel to Kenya. Fraley is at the time Monsanto's chief technology officer, credited as being an inventor of genetically modified seeds and one of their most avid proponents (after the 2018 acquisition, Bayer would employ him as a consultant). Monsanto has donated "our most advanced breeding technology and our best genetics," so that WEMA scientists can develop seeds specifically targeted for their soil and climates, Fraley adds, stressing that Monsanto doesn't profit financially from the endeavor. "We've created basically the largest public breeding program for climate-smart corn in the world."

Mariam Mayet, director of the Johannesburg-based African Centre for Biodiversity, bristles at this claim. Mayet says WEMA and Monsanto are engaging in "profiteering disguised as philanthropy," and her organization is waging a lawsuit against WEMA on these grounds.

The WEMA program, which reaches not just Navakholo but about two hundred thousand rural farmers in six countries throughout central and eastern Africa, is controversial for many reasons, not least because it's priming these rural communities to become markets for genetically modified seeds. While DroughtTego seeds are gene-altered, they are not technically GMOs (a distinction I'll explore in a moment). WEMA is also testing a GMO version of Tego known as DroughtTela at several research stations near Navakholo, with the hope of improving drought tolerance and pest resistance in the crop.

The role of GMOs in agriculture has been hotly debated in Kenya. The government here banned both the import and commercial cultivation of GMO crops in 2012, citing concerns about their potential environmental and human-health impacts. Of the fifty-four countries on the continent, only seven—South Africa, Egypt, Nigeria, and Ethiopia among them—allow commercial cultivation of GMOs, mostly cotton grown for textiles. Yet the restrictions are under fire: a growing number of Western and African scientists and politicians have railed against the anti-GMO laws, arguing that genetically engineered and edited seeds can help African countries become agriculturally self-sufficient and help alleviate the growing stresses of heat, drought, and invasive pests.

I arrived in Kenya aligned with the skeptics. As with many Westerners, the mention of GMOs and Monsanto instantly raised my hackles, though I had no scientific basis for my concerns. To me GMOs represented the worst aspects of industrialized agriculture in the United States, but I also wondered whether, in an era of climate volatility and rapid population growth, bioengineered seeds could benefit rural African farms and the people who tend them. It seemed possible that rising pressures might justify the adoption of controversial tools.

ROBB FRALEY IS sitting in his desk chair holding a glass box. His office is more modest than I'd expected for Monsanto top brass—low ceilings, fluorescent lights, venetian blinds askew, clutter and tchotchkes everywhere. The dozens of stacked books and DVDs on his desk include titles like *Altered Genes, Twisted Truth,* and *GMO-OMG!* ("I read every attack on genetic engineering I can find," he tells me.) The walls and shelves of Fraley's office hold decades of framed pictures of his wife and three kids, along with photos and letters from his mentor, Norman Borlaug. Resting on a reading chair, inexplicably, is a stuffed metallic pea pod. On the table next to it is the National Medal of Technology Fraley was awarded by President Bill Clinton. Fraley is also a World Food Prize laureate, and in 2017 he joined with two other recipients of this award—Akinwumi Adesina, president of the African Development Bank, and Pedro Sanchez, a soil scientist with the University of Florida's Institute of Food and Agricultural Sciences—to call for global action to "save Africa's crops" from an increasing pest epidemic with GMO seeds engineered for insect resistance. The invasive pest, a caterpillar known as autumn armyworm, was beginning to ravage tens of thousands of acres of Kenya's maize crops, and chemicals were failing to contain the problem. Oniang'o, who won the Africa Food Prize in 2017, voiced support for their campaign.

"This is the power of genetics," Fraley tells me, tapping the lid of the glass box. Inside, on the left, there's a specimen that looks like a small brown petrified Twizzler; on the right, a yellow, shucked twelve-inch ear of Bt-corn. "The shrunken thing is teosinte, an ancient wild grass that's the mother plant, the progenitor, of maize," he says. "What's amazing is there are only about five or six genetic changes responsible for the differences between teosinte and modern strains of maize. A half-dozen mutations over a few thousand years took us from here to there," he says, sliding his finger left to right.

Fraley launches into an impromptu seminar on plant-breeding history. Humans have been fiddling with plant genomes for thousands of years, deliberately and inadvertently, he tells me: before farmers came along, tomatoes were small and bitter, carrots were runty and pale, grapes were pea-sized and grew in sparse clusters, and leafy greens "were so sour they'd make your lips curl." Fraley makes the case that just as food production is hardwired into who we are—it fueled the growth of civilization—humans are also hardwired into food.

Much of the early genomic manipulation of grains may have been accidental. An essay in *Nature* by Jared Diamond, "Evolution, Consequences and Future of Plant and Animal Domestication," describes the early evolution of wheat and barley, which originally had seeds on top of their stalks that abruptly shattered, making them hard for humans to gather. Then along came a random single-gene mutation that prevented shattering: this, wrote Diamond, "is lethal in the wild (because the seeds fail to drop), but conveniently concentrates the seeds for human gatherers. Once people started harvesting those wild cereal seeds, bringing them back to camp, accidentally spilling some, and eventually planting others, seeds with a non-shattering mutation became unconsciously selected for, rather than against." So it went for thousands of years—farmers selecting mutation after favorable mutation and along the way producing ever bigger, softer grains, less bitter vegetables, and fatter, sweeter fruits.

The story of maize is intriguing because the plant's origins were a genetic mystery until the 1930s. Modern maize doesn't grow in the wild anywhere on earth. Scientists assumed its ancestor was extinct until the Nobel-winning geneticist George Beadle discovered in 1934 that drastic mutations over a relatively short period of time had transformed teosinte into the caloric titan of modern agriculture. Just a handful of key mutations improved the types and amounts of starch production within the grains, along with kernel size, shape, and color as well as the length and number of cob rows. Those mutations had also improved the plant's ability to grow in different types of soil and different climates, and to develop resistance to certain pests.

What early maize plants had going for them was extremely high levels of fertility. "Corn loves to prosper," Fraley tells me, "and because it's so naturally prolific, it's been able to adapt to just about every climate that humans have adapted to, including tropical and temperate, dry and rainy, cool and warm." Which means there's a huge gene pool to choose from when scientists are developing new varietals. It explains why conventional—which is to say, hybrid or non-GMO—breeding of maize has been so successful. The crossbreeding techniques pioneered by Mendel, Darwin, and later Borlaug enabled scientists to more efficiently identify and control favorable mutations among the vast range of corn types around the world to produce higher-yielding crops.

Fraley points out there are important distinctions between the corn grown in Kenya and that grown in America. Whereas corn is mass-produced in the United States for many non-food applications like ethanol, maize is beloved throughout Africa for good reason: "It's a caloric dynamo," says Fraley. "This grain grows more calories on less land than any other crop on earth." *Washington Post* columnist Tamar Haspel made this argument in "In Defense of Corn," an article that laid out a calorie-per-acre comparison of major crops. Modern corn seeds produce roughly fifteen million calories per acre. Potatoes yield close to that, and rice comes in at eleven million calories per acre. Soy crops produce six million and wheat about four million. "Corn's caloric advantages," says Haspel, "are critical to the subsistence farmer."

THERE ARE LIMITS to what can be done with conventional breeding. Modern genetics has enabled scientists to speed up the breeding process of new maize varietals using tools Mendel and Darwin could have only imagined. DroughtTego seeds were developed, for example, with "marker-assisted" breeding, a method that's considered conventional, yet from a technology standpoint it's hardly humdrum. The method allows for precise changes in a seed's genome, including the insertion of new genes encoding desired traits from the same species. The process is valuable for its speed, significantly reducing the time it takes to produce new varietals—an advantage when crops need to rapidly adapt to changing conditions like invasive insects.

Here's the key distinction between GMOs and conventional seeds: whereas conventional crops acquire traits from the same or a similar species of plant, GMOs can get their new traits from a different organism. In lab experiments, for example, rat genes have been inserted into lettuce to make the plants produce vitamin C, and genes from the cecropia moth have been spliced into apple trees to protect against fire blight disease.

DroughtTela, the GMO version of Tego, contains two foreign genes, one from the bacterium *Bacillus subtilis,* which is present in both soil and human gastrointestinal tracts and in lab experiments has proven to help plants manage water more efficiently as they grow. The second gene, derived from another common soil bacterium, *Bacillus thuringiensis* (or *Bt*), enables the plant to internally produce its own organic insecticide. Not surprisingly, these crops have counterparts in the United States: DroughtTela is a variation on DroughtGard, a Monsanto seed already grown on three million acres of American farmland. No less than 90 percent of the corn grown in America (about eighty million acres) is genetically modified with traits including *Bt*—a reality often cited in criticisms of the U.S. corn industry, known to its foes as King Corn.

Fraley argues that corn engineered to be inherently drought-tolerant and pest-resistant will benefit the poorest populations in the driest climates above all others. Sylvester Oikeh, the Nairobi-based scientific director of WEMA, supports Fraley's claim that Monsanto and its parent company, Bayer, don't stand to gain financially from this endeavor in the near term. "The company has donated its breeding technology to WEMA royalty-free, and has volunteered to train its scientists to breed and produce the seeds locally," Oikeh says. "This is not a for-profit effort." That's true for now, but eventually Monsanto will benefit from creating new markets for its seeds. WEMA scientists have developed more than one hundred varietals for different soil types and regions, all sold with the DroughtTego brand. "Seeing the Tego brand is sort of like 'Intel Inside,'" Bayer's Mark Edge tells me. "The farmers begin to associate it with high yields and resilience." For now Monsanto doesn't

Mounds of ugali *at the center of our feast in Navakholo*

charge a premium—smallholder farmers are paying "the standard prices of the local seed markets for advanced seed technology," says Edge—but as the seed markets mature, Monsanto will begin to sell the seeds for a higher profit.

Mariam Mayet, of the African Centre for Biodiversity, argues that Monsanto is engaged in a devious form of agricultural imperialism—playing up short-term humanitarian ambitions when it's really a long-term effort to capture and control new markets. The WEMA program, she says, "is really the scam of the century. They're pushing a technology-driven, one-size-fits-all approach to food production focused on monocultures and GMO seeds, which fails to utilize the traditional knowledge and skills of the farmers."

Kenya's debate over GMOs is now at a fever pitch. More than one hundred environmental groups in Africa, many of them affiliates of Western organizations like Friends of the Earth and Greenpeace, have rallied in support of restrictions against GMOs and stoked fears that GMOs are dangerous to human health and a threat to the purity of Kenya's local and indigenous crops. For years, government officials have echoed the concerns of these activists, but in January 2019, the country's National Environment Management Authority (an organization similar to the U.S. Environmental Protection Agency) approved the introduction of the first commercial GMO crops on Kenyan soil.

I WAS NINETEEN when the first GMO corn was introduced in the United States, and probably soon thereafter I unknowingly ate my first GMO corn chip. I now have more than two decades of GMO crops in my system. There's a high probability that you have ingested likewise more GMO food than you realize. Seventy percent of processed American foods, including pizza, crisps, cookies, ice cream, salad dressing, corn syrup, and baking powder, now has at least one genetically engineered ingredient. Indeed, millions of Americans have been eating GMOs since the mid-1990s, most of us unwittingly. This reflexively irks us, even if we have no internal or external evidence that the foods have affected our health in a negative way.

More than 400 million acres of GMO crops—mostly rice, corn, cotton, and soybeans—are currently growing on farms in thirteen countries worldwide, among them Argentina, Brazil, Canada, China, Australia, Germany, and Spain. The United States is responsible for more of these acres—all told, about 180 million—than the others put together. Critics argue that GMO adoption has been dangerously rash. "Manufacturers of genetically altered foods have exposed us to one of the largest uncontrolled experiments in modern history," asserted Dr. Martha Herbert, a pediatric neurologist and public health advocate.

Yet every major national scientific society, including the National Academy of Sciences and the World Health Organization, has concluded that the GMOs on the market pose no human health threat. Despite efforts by opponents to link the technology to health problems ranging from allergies to cancer, "these claims are not based on science," says Pamela Ronald, a plant geneticist who runs a large laboratory at the University of California, Davis. The only major scientific study to connect GMOs to cancer—by a team led by Gilles-Eric Seralini, a French scientist who proposed that GMOs were linked to tumor growth in rats—was officially retracted by the journal that first published it, *Food and Chemical Toxicology*, on the grounds that it contained inconclusive data.

"There's just something about genes that terrifies people, some deep fear that we're tinkering with the essence of life, that we're going against nature," says Ronald. When GMOs were first introduced, there wasn't much evidence proving its safety, but that has changed. "Since genetic engineering techniques were first commercialized in the 1970s, there has not been a single instance of harm to human health or the environment," says Ronald. "After over twenty years of careful study and rigorous peer review by thousands of independent scientists, *every* major scientific organization in the world has concluded that the genetically engineered crops currently on the market are safe to eat."

Even some of the most outspoken critics of the technology, including the food writer Michael Pollan, say that it doesn't necessarily pose a greater health threat than other forms of plant breeding. "I haven't read anything to convince me that there are *inherent* problems with the technology. I think most of the problems arise from the way we're choosing to apply it," Pollan tells me.

Mark Lynas, who was an early GMO activist in England, so loathed the idea of manipulating plant genomes that he destroyed acres of GMO corn crops in the middle of the night. In 2018, he published a mea-culpa book, *Seeds of Science: Why We Got It So Wrong on GMOs*, which asserts that the scientific evidence for the safety of GMOs is as robust as the science behind climate change. He writes, "I couldn't deny the scientific consensus on GMOs while insisting on strict adherence to the one on climate change, and still call myself a science writer." Haspel of the *Washington Post* has also advocated for a shift in public perception around GMOs. "The argument against GMOs has never really been about the GMOs themselves," she says. "It is about a corporate-dominated, industrialized food system for which GMOs serve as a kind of proxy."

The first applications of genetic engineering were in medicine. In 1972, scientists began engineering enzymes from yeast and bacteria to create lifesaving drugs, including insulin, and, eventually, cancer treatments. To this day, insulin treatments for diabetics are derived from GMOs. It wasn't until the mid-1990s that agronomists began applying the methods to crop science for consumer use. First introduced in 1994, GMO crops spread fast. "Farmers went from growing half a million hectares of GMOs to fifty million in one decade," Fraley tells me, "the fastest adoption of any innovation in the history of agriculture."

The new gene-editing tool CRISPR could eventually make the modification of a plant's or animal's genome nearly as simple as manipulating a file in Photoshop. "It's cheap where GMO is expensive, so cheap that you can buy a kit to CRISPR a bacterium right at home for $159. The biotech industry estimates that it takes $130 million to bring a genetically modified crop to market," writes Haspel. And because it's cheap, it's becoming as accessible to academic researchers as it is to corporate giants. In 2017, scientists used CRISPR to erase HIV from laboratory animals, and breed piglets with DNA scrubbed of a lethal virus, making it safer to raise pigs for human heart transplants. But while these medical achievements were widely celebrated, recent CRISPR applications in the food space were not. Academic researchers have developed gene-edited mushrooms and potatoes, for example, that don't brown when exposed to oxygen and are therefore less prone to waste—these raised suspicions.

The very good reason for this suspicion is that most mainstream applications of genetic engineering in agriculture have been seriously flawed. They're designed to benefit corporations more than consumers, says Pollan, "enhancing the productivity of industrial farming, increasing herbicide sales, and reinforcing practices like monocropping."

Monsanto's Roundup Ready seeds—the crop that has become nearly synonymous with GMOs—are engineered with special "tolerance" genes that help them withstand the spraying of chemicals that kill virtually every other kind of plant. These chemically tolerant plants now account for 90 percent of all the corn, cotton, and soybeans in the United States. Many of the products have backfired. Farms using Roundup Ready and other herbicide-tolerant seeds have developed "superweeds" resistant to the chemicals, prompting farmers to use more and stronger weed-killing sprays. On top of this, a subsidiary scientific body of WHO has recently deemed glyphosate, the key ingredient in Roundup, potentially toxic to human health. None of this has promoted much public confidence in GMOs.

"The major GMO crops are failing," Pollan tells me. Many not-so-major GMO crops have also failed: the FlavrSavr tomato, for example, which was engineered in 1994 to be slow-ripening, rot-resistant, and boldly flavored, was yanked in 1997, three years after it was put on the market, due to low demand and high production cost. Golden Rice has been a more painful flop, in part because it began with such high hopes. Geneticists thought that rice engineered with high levels of beta-carotene could cure the epidemic of childhood blindness caused by severe vitamin A deficiencies in many food-insecure countries. Yet after a decade of development, they haven't been able to crank up the beta-carotene levels high enough to relieve the crisis. The most recent effort to make a popular GMO—the so-called Arctic apple, which entered American supermarkets in 2017—has also been criticized. Geneticists have "switched off" a gene sequence that causes the flesh of Golden Delicious apples to turn brown when cut and exposed to oxygen. More kids will eat the sliced apples in their lunch boxes and less food will be wasted, the creators hope. But while the list of GMO items in American supermarkets is growing—DelMonte, for example, has gotten FDA approval for a pineapple genetically modified to have pink-colored ("rosé") flesh—the products have so far gained neither trust nor traction among mainstream consumers.

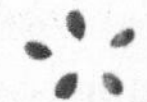

GIVEN THE FAILURES and frivolities of GMO products currently on the market, it's not surprising that food with the "non-GMO" label jumped from $8 billion in 2015 to $26 billion in 2018. Pew polls say that about half of Americans on both sides of the aisle believe GMOs are worse for their health than non-GMOs, even if there is no science to support this. Trader Joe's and Chipotle, among many others, have sworn off using and selling GMO ingredients, as have brands including Dannon and Triscuit. "The companies that produce [non-GMO] brands are guilty of crimes against rationality," observed Michael Gerson, another *Washington Post* columnist, proposing a boycott *against* the boycott on GMOs. The labeling trend has escalated to the point where there are now products marked non-GMO—including salt, candles, and even kitty litter—even though they're made of inert ingredients that don't possess a genome that could've been modified.

Pamela Ronald is one of many scientists who believe that the public backlash against GMOs ignores key successes of the technology. Take Bt-cotton, for instance, which has cut the amount of chemical insecticides applied to crops globally by millions of pounds a year. *Bt*—the bacterium also spliced into the DroughtTela product—is used in a spray formulation by organic food producers worldwide, and is no more toxic to humans and animals than table salt. The USDA has reported a tenfold reduction in the use of insecticides as a result of the engineered *Bt* trait.

Ronald also cites the example of papayas genetically engineered to resist ringspot virus, which have helped save the Hawaiian papaya industry, and "scuba rice," a strain of flood-tolerant rice that she and her colleagues bred to grow in submerged fields using marker-assisted breeding. In 2017, more than six million subsistence farmers grew this rice in flood zones of Bangladesh and India. "It's a shame to demonize an entire technology because of Roundup Ready," she tells me.

Ronald points out that the "public paranoia around GMOs," as she calls it, echoes the response people had in the late nineteenth century around advances in hybrid breeding. "Many worthy people objected to the production of hybrids on the ground that it was an impious interference with the laws of Nature," wrote Maxwell Masters, a British taxonomist and botanist, in 1899. It took several decades for the public to acquiesce to this technology . . . just in time for James Watson, Francis Crick, and Rosalind Franklin to discover in 1953 that DNA is the molecular basis for living organisms, throwing open the gates of genomic science. "It is an awesome power to be able to decode and reorder genes," Ronald tells me, "to move them between living things. Aspects of it are scary, sure—but it's also an opportunity to do a world of good." Nathanael Johnson, a columnist at the environmental website Grist.org who spent months investigating GMO crops, tends to agree: "Trying new things can be risky. Not trying new things—staying on our current trajectory—is even more risky."

Hassan Adamu, Nigeria's minister of agriculture and rural development, has warned that the fears around GMOs could disproportionately hurt climate-vulnerable African populations. "It's possible to kill someone with kindness, literally—that may be the result of well-meaning but extremely misguided attempts by European and North American groups that are advising Africans to be wary of agricultural biotechnology," he says. "If we take their alarmist warnings to heart, millions of Africans will suffer and possibly die." An op-ed by a Zimbabwean scientist in the *Wall Street Journal* echoed this sentiment, railing against her nation's refusal to accept genetically modified crops during a widespread famine: "My country's government would rather see people starve than let them eat genetically modified food. . . . The rejection of GMO food aid is a humanitarian outrage—a manmade disaster built on top of a natural disaster."

Ruth Oniang'o's perspective on Monsanto and GMOs is measured. "This is not a question of good or evil," she says. "All the science I've seen shows that the benefits of GMOs far outweigh the risks, and the risks can be controlled." She notes that the resistance to Monsanto comes "from people who can afford to pay a premium for their foods" and who romanticize an agricultural past that can't meet the environmental and population pressures ahead. "We don't have that luxury in Kenya. We are transitioning from beggars of food to exporters of food. There can be no progress for people who cannot feed themselves."

Oniang'o supports, like Annie and Chris Newman, a third way approach to agriculture that moves beyond the binary thinking of ag companies and sustainability activists in the United States. Her vision for food production in Africa blends strategies of the past and present. "The way you discuss agriculture in America is as though there are two paths—either old-world agroecology or high-tech agribusiness," she explains. "Why can't these approaches coexist? They must coexist. Our population is expected to double in the next three decades. We need an agricultural sector that keeps up with the pressures, rather than one that keeps on falling behind. We need both indigenous vegetables and modern seeds, nutritional diversity and high-yielding grains."

THE SCIENTIFIC RESEARCH station in the Kenyan town of Kitale, just north of Navakholo, resembles a small college campus. Rambling lawns with bristly yellow grass surround several low-slung cement buildings containing laboratories, offices, and dormitories. In a corner of this campus is a patch of farmland about half the size of a football field bordered by a chain-link fence with a coil of barbed wire on top. Inside the fence is a rectangular grid of thousands of towering maize stalks glittering green against the dun-colored surroundings. About half of the crops are variants of Monsanto's controversial GMO crop, Bt-corn. Dr. Dickson Liyago, who heads up the maize-breeding program for WEMA, has brought four scientists from his team to collect data for their study-in-progress. Bt-corn was originally developed to deter the European stem borer, and the scientists are testing its efficacy against the moth's pernicious cousin, the African stem borer, along with the autumn armyworm, which is decimating maize crops across the continent. They're measuring the efficacy of the Bt-maize

At the GMO test field with Dr. Liyago (center) and his team

seeds against the leading maize seeds on the market, which have conventional pest-resistance properties. When they complete the Bt-maize research, Liyago will move into phase two of the research, replacing the plants in this test field with DroughtTela maize that contains "stacked" traits for both pest resistance and drought tolerance.

Liyago unbolts two padlocks fastened to the metal chains looped around the front gate of the chain-link fences and swings it open. We enter the test plot, passing signs that read BIOHAZARD and AUTHORIZED PERSONNEL ONLY, and step into a small shed where we suit up in long kelly-green laboratory coats (if pollen from the test crops collects on our coats, it'll be visible against the green background and more easily removed). We take turns stepping into a shallow tray of antibacterial liquid to disinfect our shoes from pollen we might be carrying by foot into the research area.

Liyago leads us into the labyrinthine cornfield. The maize stalks in their narrow, quiet, tunnellike rows are so green and healthy they seem to glow. They stretch so high above us the sky is barely visible. It feels at once like bushwhacking through a jungle and like walking through library stacks. The plants in every plot and row are meticulously spaced, ordered, and enumerated, each one with a sign describing its genetic pedigree. A "border crop" of five outer rows of maize surrounds the test area, helping protect against the inflow and outflow of wind-borne pollen. Within the test area there are six contiguous rectangular plots, each with a different variety of maize seeds that contain the *Bt* trait. Interspersed within the rows of GMO plants are the same seed varieties, but lacking the *Bt* trait, as well as rows of the leading maize varieties currently on the market. The researchers call these plants "commercial checks."

"I have a test for Amanda!" Dr. Liyago shouts. He's a short, reedy man in his seventies with metal-rimmed glasses and a broad smile that shows every well-aligned tooth in his mouth, including the hindmost molars. "Tell us, which among these plants is genetically modified?" he asks. I falter, still thinking about bushwhacking and book stacks. "Walk down the row and look closely," Liyago urges. I begin to see that every three rows, the leaves on the maize are riddled with tiny ragged holes, like fabric that's been blasted with shrapnel. In the neighboring three rows, the maize leaves have no holes to speak of. "These?" I venture, pointing at some of the hole-free plants. "*Nzuri!*" says Liyago. "You are right. Look how robust they are—strong growth characteristics and vigor. But the adjacent neighbors are suffering. The Swiss-cheese leaves have clearly been attacked!"

Liyago's colleague Dr. Omar Odongo rips open a corn husk on the Bt-maize and brushes away the silk to show the rows of perfect, pearly kernels. Then he shucks a cob of a non-*Bt* plant and the kernels are missing or misshapen; there's a gooey brown patch where a fat gray caterpillar rests, days away from morphing into a moth. "The presence of the gene in the *Bt* plant is like an on-off switch for the pest," says Odongo. "Larvae nibble a few bites of leaf and die." The autumn armyworm is native to the Americas and didn't make its way to Kenya until 2016. It's now in more than thirty African countries and spreading rapidly. The caterpillar has caused billions in losses of maize, sorghum, and other staple crops on the African continent since the beginning of 2017—a big-deal hit for farmers struggling to survive.

Most conventional methods for controlling these pests fall short.

Those who can afford pesticides use them, such as fenthion, an organophosphate that may be toxic to the human nervous system. They'd do far better to apply the chemical version of *Bt*, yet it's too expensive for smallholder farmers. The cheaper chemicals like fenthion are used in high doses to reach the innermost recesses of the leaves and stalks. In the case of the stem borer, the female moth lays about two hundred eggs at a time, embedding them deep at the base of the maize leaves where the cobs form. The larvae hatch in days, feeding on the leaves and then tunneling into stalks and cobs, where they grow and pupate. Farmers who can't afford chemicals often attempt to manually sprinkle ash or sand, a pinch at a time, into the whorls of every leaf on their young corn plants—tens of thousands of leaves in a one-acre farm—hoping that the barriers will kill the larvae and prevent them from entering the stalks. That painstaking process rarely works.

"The results of the trials so far are good," Liyago tells me. "The Bt-maize is producing yields about 40 percent higher than the non-Bt plants." It's an example, he adds, of achieving a result with a GMO that you can't get to with a conventional crop. The transgene saves the small farmer money and time, cuts the use of toxic chemicals, and increases yields, food security, and farmer income, says Liyago. When I ask him about the negative impacts of Bt-maize pollen on beneficial insects like butterflies, and of the threat of genetic drift, he waves it off, saying that fifteen years of testing has shown that the corn doesn't contain enough toxin to harm beneficial insects. I corroborated his take with reporting in the science journal *Nature*. (Jane Rissler, with the Union of Concerned Scientists, told *Nature*, "We are pleased it looks like transgenic corn pollen is harmless.")

While Liyago calls Bt-maize "an economic and environmental win" for Kenya, he's more cautious about the potential benefits of Monsanto's drought-tolerance trait. If you've ever tended a house plant, you may have noticed that some are forgiving when you forget to water them—ferns, ivies, and succulents, for example, can perk back up when they get a drink—while others are less tolerant. The reasons for a plant's resilience to water scarcity are hard to understand. Since the midseventies, companies and institutions have poured billions of dollars and decades of research into water efficiency and drought tolerance in plants, and the only definitive thing they can agree on is that "it's really complicated," Mark Edge of Bayer tells me.

Marcia Ishii-Eiteman, a senior scientist at Pesticide Action Network and a vocal critic of GMOs, says it galls her that GMOs are being pushed as a possible solution to drought when there's still little evidence that drought tolerance can be engineered into plants: "It's just dishonest." She references Jian-Kang Zhu, a molecular geneticist at Purdue University, who said, "Drought stress is as complicated and difficult to plant biology as cancer is to mammalian biology."

There's no single gene that governs the way plants respond to drought stress, but rather "a whole complex suite of them, and they differ from plant to plant," says Edge. He admits that the in-the-field results of Monsanto's DroughtGard seeds (the U.S. equivalent of Tela) have been mixed. During some drought events in the American Northwest, the crop has performed very well, yet on midwestern farms with similar water shortages, it hasn't. He isn't sure exactly why.

To solve the mystery of drought tolerance in plants, scientists have to consider the stages of development in which the plants need the most water. "We know that if drought strikes maize within two weeks before the flowering stage, it can suspend pollen development," Liyago tells me. "Within two weeks after the flowering stage, it can retard grain development. Even if you bring water back after those critical periods, most plants can't recover." They also consider the mechanisms plants use to draw water up from the soil. Longer roots can tap deeper water reserves; wider and more numerous circulatory vessels can be more efficient at carrying water from stem to leaves. Photosynthesis is also a critical factor: when plant leaves open their stomata to take in carbon dioxide, they also release water vapor as part of a natural cooling process.

Understanding the methods that plants use to survive environmental stresses is "the next big frontier" for botanists, says Pamela Ronald. She and her team have spent five years trying to develop drought-tolerant crops. The challenge is so complex and the research so costly, they're hard to tackle without an R&D budget the size of Bayer's or Syngenta's. Yet a growing vanguard of university and government scientists is entering the fray. At the University of Cape Town in South Africa, researchers are studying *Myrothamnus flabellifolius,* a so-called resurrection plant that can bounce back from near total water deprivation. The plant can lose up to 95 percent of its water—less than the water contained in a seed—and enter a state of dormancy or hibernation for months or even decades, then spring to life when rains return. Through genetic modification, the researchers hope to bring this miraculous skill to teff, a native African grain that's high in protein. Scientists at Technion University in Israel have meanwhile successfully engineered similar "resurrection" genes into tobacco.

Elsewhere, scientists in Argentina have developed a soy plant spliced with genes from a naturally drought-tolerant sunflower, and recently gained approval from the Argentinian government for commercial cultivation. At the Oak Ridge National Laboratory in Tennessee, scientist Xiaohan Yang is studying the way plants like the agave cactus store and manage water, in hopes of developing crops with these abilities. If he succeeds, vast tracts of desert could become productive farmland.

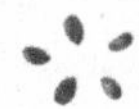

AS RUTH ONIANG'O sees it, this wave of new research tells a hopeful story of climate pressures driving well-intentioned innovation. "The world does not stay stagnant," she tells me. "We just cannot stand still. The human mind must always be moving and pushing at the edges of possibility in every field, medicine, communications, engineering, and transportation—that's also true for agriculture."

The day before I leave Nairobi to head home, I visit the farm of Mary Matete, about fifteen miles from the farm in Emuleche where Ruth was raised. Mary, forty-one, is recently the founder of the New Technology Group, a collective of neighboring women who meet regularly to discuss modern farming practices. Mary and her husband, Robert, who is sixty-four, have nine kids, ranging in age from five to twenty-two. Mary does most of the labor on their 1.5-acre farm, where she now grows a combination of DroughtTego and soy crops, rotating them regularly to keep the soil healthy. She reserves a quarter acre of their land for growing groundnuts, cabbages, onions, nightshades, and yams.

Mary joined ROP in 2012 after two seasons of failed sugarcane crops. The group helped her revive her soil with fertilizer and then converted her crop to Tego maize. In the four years since she joined ROP, her yields have increased more than 400 percent, from 18 bushels of maize a year to 74 bushels. The success drove her to establish the New Technology Group and collaborate with her neighbors. She and sixteen of her neighbors have worked together to establish a line of credit with local suppliers, collectively store their surplus, and use their purchasing power to negotiate prices on seeds and other supplies.

Mary Matete

It's progress—but not enough. Mary and Robert's 74-bushel yield is still only a third of the 200-bushel yield that's reaped on the average Iowan farm. Mary grossed the equivalent of $990 in 2016, but invested more than three-quarters of her income into seed, fertilizer, and storage bags for the following season. She netted the equivalent of about $180 to support their family of eleven for a year.

A few hours into my visit to the Matete farm, just before noon, two of Mary's children, thirteen-year-old Jane Beth Aswani and nine-year-old Jonas Akweneno, come home from school, having been turned away for delinquent payments on tuition. Mary is upset. I ask about her ambitions for her children—will they carry on the family farm? Mary shrugs. The future is bigger farms and modern agriculture, she tells me. This will liberate small farmers to enter new economic sectors. She notes that Kenya, like so many African nations, is moving toward an increasingly technology-driven economy. She's heard that Google and Microsoft have recently established offices in Nairobi, which has dubbed itself "Silicon Savannah." But farms still produce about a third of the country's GDP and employ thirty million citizens. Whatever direction her kids go, in or beyond agriculture, says Mary, education and some amount of tech savvy will factor into their success.

During a pause in our discussion, Robert takes my hand and says, "If, through the almighty God, you are sympathetic, do nothing but take one."

"Take one?"

"A child," he says.

Too much time passes before I understand that Robert has asked me to raise one of their children in the United States. I mumble a response about being honored but "it would be difficult to arrange." Then I'm rummaging through my backpack for money. I ask what the cost is for a school semester. I dig out two thousand shillings (the equivalent of twenty dollars) to cover a year of schooling for Jane and Jonas and hand the money to Mary. She's visibly embarrassed and says nothing. Robert kneels on the ground and recites a prayer.

Only later, on the way to the airport, do I realize how much I've underestimated both the farmers I've met and the risks they face. I am ashamed—about my offering to the Matetes and its futility. About the fact that it's taken me this long to recognize the flawed premise on which I'd based my reporting—for one thing, that Americans should be concerned about which modern practices are adopted on African farms. The farmers I'd met were discerning practitioners, not victims, of new farming technologies, and no less capable of judging their costs and benefits than anyone else. They live closer to the land in much greater numbers than we do, which means they inherently possess a deeper commitment to sustainability. They also have far more to gain from technology than we do: resilience against the pressures of climate change, to which they are more vulnerable than perhaps any other population on earth, along with freedom from the drudgery of low-yield growing practices that American farmers have been enjoying for more than a century. "It's easy for Westerners to say let's go back to the way things were in agriculture," Ruth tells me. "But Africa is trying to move away from the way things were. And to do that, we need to consider every available tool."

CHAPTER 4

RoboCrop

He made a mistake. Now it's time to erase that mistake.
—DICK JONES IN *ROBOCOP*

THE MOST IMPORTANT thing I learn in Kenya is circumspection. I begin to see that at least a few of my fears about modern food production are, if not irrational, then misbegotten and unsupported by reliable scientific evidence. This may be true for a lot of us. We fear the human health impacts of GMOs even as every major scientific organization has concluded that genetic engineering is inherently no more dangerous than other methods of crop breeding. We assume that GMOs are inextricably bound up with all the other scourges of corporate-dominated, industrial agribusiness—from monocropping and the rising obesity epidemic to algae blooms and the decline of arable soil. Most of us, if we're honest, don't know where to begin parsing the science behind these problems or how to prioritize among them.

Ruth Oniang'o tells me that while she sees promise for GMOs and advanced seed breeding, she has serious concerns about the growing volume of industrial fertilizers and pesticides that are entering Kenya's food system. Lately, Cessna airplanes have been spraying clouds of chemicals on expanding tracts of industrial and government-run farmland. The algae blooms in Kenya's lakes and along its coastline are spreading. Ruth fears the consequences of these chemicals on soil quality, water purity, and human health. Her concerns resonate with me because I also worry about—more than anything else, really (with the possible exception of threats to my coffee supply)—the chemicals used in agriculture.

It must have been Rachel Carson's *Silent Spring* that first made me aware of the risks of pesticides. The book, which was released in 1962 and has since sold millions of copies, uses sharp prose and exacting scientific detail to expose the destructive power of DDT, an insecticide that was commonly used by American farmers during much of the twentieth century. The book helped mobilize the early stages of the modern environmental movement, the creation of the Environmental Protection Agency in 1970, and a ban on DDT in 1972. Despite the impact of *Silent Spring* and the decades of activism that followed, many other potent agriculture chemicals have since been used—sometimes with horrific consequences.

Soon after Carson's book was published, in 1967, the United States sprayed five million gallons of Agent Orange, an herbicide with the toxic chemical dioxin, across the jungles of Vietnam in an effort to defoliate the countryside so the enemy couldn't hide. The ecological and human-health consequences of this chemical campaign were lasting and grave. In 1984, an accident at the Union Carbide plant in Bhopal, India, unleashed a miasma of methyl isocyanate, a poisonous ingredient in the pesticides produced at the facility, killing an estimated 15,000 people and harming many more. In 2010, after decades of use, another controversial herbicide, atrazine, was found to be castrating male frogs and causing them to lay eggs.

Some of the problems of chemical use in large-scale agriculture have been reined in, but concerns persist. The dead zone caused by chemical fertilizer runoff into the Gulf of Mexico now covers more than 8000 square miles. Those chemical fertilizers can also evaporate into the air to form nitrous oxide, a greenhouse gas three hundred times more potent than carbon dioxide. In Iowa, contamination caused by agricultural runoff has given way to a condition known as "blue baby syndrome," in which the flow of oxygen in the bloodstream of infants is limited by nitrate residue from fertilizers that leach into tapwater. The massive die-off of bees known as colony collapse disorder has been linked to the use of neonicotinoids, a common insecticide ingredient made from nicotine-like chemicals that harm the bees' ability to reproduce. A number of recent academic studies have linked birth abnormalities with exposure in pregnancy to high levels of organophosphate pesticides, which are commonly used in agriculture and landscaping.

To be clear, the risk of exposure is far higher for farmworkers than it is for average consumers. The amount of chemical residue in our foods (including conventional produce) is extremely low, according to most toxicology research out there. Even the Environmental Working Group, a pesticide watchdog organization, has said that the health benefits of eating nonorganic fruits and vegetables far outweigh the risks of chemical residues on those foods. But the *cumulative* effect, over time, of certain agrochemicals on our ecosystems and public health, especially as food demands grow and pest populations increase and the amount and quality of arable soil decline—yes, that's worth worrying about.

Which is why I went in search of someone who could help answer this question: How will we produce affordable food for billions of people in the coming decades with a much lower volume of chemicals? Eventually, I found my way to a Peruvian-born Silicon Valley engineer who's building an army of robots designed to do exactly that.

JORGE HERAUD IS in a Californian lettuce field and he's about to lose his mind. It's a balmy, cloudless day in April 2014, and Salinas Valley stretches out around him like a Hidden Valley Ranch commercial, its endless rows of emerald lettuce leaves pushing up through the black soil. Heraud has come here to test "Potato," a robot that may be the agricultural equivalent of an Apple-1 prototype circa 1977. Its success could shape the future of lettuce farming, and of farming, period. Just as Ruth Oniang'o sees a synergy between old and new forms of agriculture, Heraud shares in this third way thinking, and Potato is evidence of that. Intelligent machines, he believes, are not at odds with sustainable food production, but a means to achieve it.

As Heraud looks on, Potato is trying to perform the seemingly simple task of thinning baby lettuce plants so the hardier ones have space to mature. If you're imagining, as I was, a C-3PO-style bipedal robot that walks into fields with pincerlike hands to do the yanking, Potato isn't that. It looks like a huge metal Pez dispenser laid sideways on a rack that's hitched to the back of a tractor. The bot "sees" the seedlings via cameras mounted to a rack. In milliseconds, it identifies the strongest ones and kills the weaklings with jets of concentrated fertilizer shot through tiny tubes and nozzles.

Or that's what it's supposed to do, but Heraud's prototype is on the fritz. Robots like controlled environments, and Potato's delicate equipment isn't responding well to the heat and dust or the tractor's vibrations. Electrical components are short-circuiting, nozzles are failing, dirt is gumming up the cooling fans, and the PCs aren't stable. All day long, about every half hour, Potato's monitors freeze into blue screens of death.

Heraud's agony deepens as the failures mount. For months, his team has been testing beta cousins of Potato, each with a salad-themed moniker: "Caesar," "Cobb," "Chicken," "Wedge," "Jell-O." All are early versions of a product they've officially named LettuceBot, which Heraud has prematurely begun leasing to farmers. Two days from now, Heraud has to face his investors at a board meeting. They've ploughed $13 million into his start-up and they want to hear that he's got it working already.

Heraud, who is forty-five, internalizes stress. Lately his skin has been breaking out into prickly rashes; he's had insomnia and punishing heartburn. LettuceBot isn't even what he'd originally pitched to them. He'd envisioned a robotic weeder that could perform far more complex tasks and radically reduce the use of agricultural chemicals worldwide. Such a machine would first disrupt the herbicide industry, which is dominated by Syngenta, Bayer, DowDuPont, and Monsanto. It would protect topsoil fertility, support climate-smart practices like "no-till" farming, save countless aquatic and amphibious species, diminish the public health problems stemming from chemical residues in our food, and clean up the world's waterways. Heraud had named the company Blue River Technology with these lofty goals in mind.

When Heraud confesses his field-test failures to the board, they don't vote to oust him as he'd feared, but challenge him to turn things around. In the coming months, he and his team of twenty engineers launch a 24/7 troubleshooting offensive they call "the surge." They take turns sleeping on cots in the closet of their Silicon Valley office. They call in husbands and wives to turn wrenches and clamp tubes. They redesign fans, build mounts, change materials, and reformulate chemi-

Jorge Heraud

cals. Heraud consumes Tums by the fistful. By late 2015, they have a glitch-free LettuceBot that can handle the elements. They expand their contracts with farmers in Salinas and in Yuma, Arizona, and build more machines. By early 2017, about a fifth of all the lettuce grown in the United States has been thinned by a LettuceBot.

Heraud and his investors are buoyed by the success, but other news excites them more. The microchip company Nvidia has released a computing platform with outsize processing power. It's designed for navigation in self-driving cars, but it also means that a farming robot, like Heraud had always envisioned, might be able to crunch a lot more data captured by mobile cameras than the current lettuce thinner. It means Heraud's team may be able to build that weeding robot he'd imagined after all. What Heraud couldn't possibly have imagined, however, as his team began to cobble together their first dream machine, is that in September 2017 the green-and-yellow tractor company, John Deere, would acquire Blue River for $305 million. And the oldest brand in farming, founded in 1837, will be on board with Heraud's grand vision—not just of slashing agrochemical applications worldwide, but of transforming food production for good.

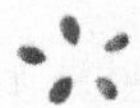

THE BLUE RIVER headquarters in Sunnyvale, California, is located in a modest, one-story building down the street from Yahoo!, Juniper Networks, and Lockheed Martin Space Systems. "Welcome to agriculture 2.0," Heraud deadpans, gesturing at the generic cubicles and gray carpet tiles that make up his office space. Of the seventy-two employees, only a few, including Heraud and his cofounder, Lee Redden, have any dirt-under-the-fingernails growing experience; the rest are software and mechanical engineers with degrees from Harvard, Stanford, Oxford, and Caltech. The "I ♥ SOIL" sticker on Heraud's IBM ThinkPad is one of few clues that this is a farming company—along with a framed photograph of a yellow Cessna crop duster like the ones Ruth Oniang'o has seen prowling above the industrial farms of Kenya.

This one's spewing glyphosate (aka Roundup) on vast Iowan cornfields. Heraud, who has deep-set blue-green eyes and a stolid demeanor, tells me he hung it there "to remind us what we want to get rid of."

Heraud grew up in Lima, Peru, a maths-loving only child of an electrical engineer and an elementary school teacher. By the age of five he was spending his free time adding columns of six-digit phone numbers in the phone book. His parents enrolled him in an international school with British instructors. Weekends and afternoons, he tagged along at his dad's company, Digita, which specialized in factory automation. In the summers, he stayed on his grandparents' farm north of Lima, where his family grew two hundred acres of tomatoes and rice.

He loved the fun parts of farm life—driving tractors and pickup trucks, raiding the sweet mango orchards, gathering eggs from the coop, and eating his grandmother's cakes and pies. But the drudgery seemed to him inane. He was up by 5:30 a.m. and in the fields with his cousins by six, pulling weeds. "I understood early on that a farm, even a small one, is basically a big outdoor factory. There'd be dozens of us kids in the fields bending and picking, bending and picking weeds. I must have been seven when I first thought, This is a repetitive job for machines."

Heraud excelled at school and by fourteen was designing software for his dad. He went to Universidad Catolica of Peru, a South American hub for mathematicians, and continued working on the side, heading up a project to automate a chicken-feed factory, creating processes for sorting, weighing, mixing, and packaging the ingredients. Soon Stanford snatched him up with a scholarship to its electrical engineering master's program. Trimble, one of the first companies developing GPS-based technology, hired Heraud after he graduated. In the midnineties, before Google-X and Tesla, he led a team that designed the first self-steering tractor. "It wasn't until we debuted the thing at a tech expo and there was a mile-long line of people wanting to try it out—that's when it hit me we'd built a better mousetrap." Self-steering tractors are now used to produce more than half of the developed world's food, says Heraud, and they've set the stage for self-driving cars.

Heraud became Trimble's director of acquisitions, buying up companies that made things like precision seeders and digital sensors for measuring soil moisture, until he realized he wanted his own enterprise. He left Trimble to get an executive MBA at Stanford, where he posted "Let's solve agriculture's biggest problems" on the university intranet. Lee Redden, a twenty-four-year-old Nebraskan robotics PhD student, replied. Redden had grown up working summers on his uncle's 6000-acre corn farm, and by fifteen, he was a professional car mechanic. He also had a thriving side business building and fixing motorcycles, ATVs, and go-carts. At Stanford, he'd cranked out dozens of robots that could perform everything from Ping-Pong training to infant CPR. "But they all just sat on a laboratory shelf collecting dust," he says. "I wanted to do something that could live in the world."

Heraud researched the scourges of agriculture: hypoxic dead zones, colony collapse of bees, human health problems tied to chemical residues in food, and the loss of topsoil. "Everything tied back to the overuse of chemicals," says Heraud. He and Redden figured they could teach machines to differentiate between crops and weeds and then eliminate the weeds mechanically or with targeted doses of nontoxic substances.

The pair first considered hot foam, laser beams, electric currents, and boiling water as weapons against weeds. They planned to market the robot to organic farmers, who spend heavily on chemical-free weeding methods, including mechanical tillage, which can be both fuel-intensive and damaging to soil. But after months of research, they faced a disappointing truth: there was no way around herbicides. "Turns out zapping weeds with electricity or hot liquid requires far more time and energy than chemicals—and it isn't guaranteed to work," Heraud says. Those methods might eliminate the visible part of a weed, but not the root. And pulling weeds with mechanical pincers is a far more time-intensive task for a robot than delivering microsquirts of poison. Heraud and Redden homed in on a strategy: "applying chemicals with radical precision; we just had to figure out how."

There was far more to the challenge, of course. This was David vs.

Goliath: two idealistic dorks proposing to upend the $28 billion herbicide industry, and, beyond that, the $250 billion agrochemical industry. There were also personal stakes: Redden would have to scrap his PhD program and forfeit hard-earned scholarship money. Heraud, who had little kids, would face years without a salary and sacrifice the executive position Trimble was reserving for him. "The only thing we knew as a certainty in the beginning," he says, "was that not working on this problem would be soul crushing."

IN THE EARLY days of Blue River, Heraud realized it was better to involve than avoid the competition. He pitched his company to the investment divisions of Monsanto and Syngenta, the very giants of the industry he was planning to eradicate—or at least thin—with his weeding robot. He wanted access to their chemists and botanists and the credibility the association would give him among mainstream farmers who could field-test his prototypes.

At first, the response was tepid. "We loved Jorge's Trimble background—smart guy—but there was some starry-eyed idealism," says Syngenta's investment director Gabriel Wilmoth, who passed on the first investment round but kept tracking the company. When he saw the LettuceBots succeed and heard the news of the Nvidia chips, he wanted in. Kiersten Stead, an investment director at Monsanto Growth Ventures, also ponied up some funding. The backing was nominal by Big Ag standards—a few million dollars—and partly a way to keep tabs on the young competition. The investments were also, you might say, an admission of defeat. "The herbicide industry faced the humbling reality that their chemists were being out-thunk by weeds," says Willy Pell, an electrical engineer and one of Heraud's first employees.

Often dismissed as the deplorables of the plant kingdom, weeds are in fact elegant masters of adaptation and procreative prowess. One dandelion, for example, can produce about 170 seeds, each smaller than the dot on this *i* and rigged with a feathery parachute that can carry it many miles before it lands and roots. By this method, over thirty million years, dandelions came to inhabit six of the world's seven continents. Jewelweed has another ingenious propagation strategy. It stores its genetic information in a "ballistic seed pod" designed to explode at peak fertility. The seeds of devil's claw, like those of burdock and cocklebur, have grips and bristles that cling to animal hooves and fur, so as to be distributed far and wide. Barnyard grass uses biomimicry to look exactly like a rice plant—its appearance and growth characteristics so similar to rice that it can outwit a seasoned harvester.

But the Genghis Khan of weeds—the one most hell-bent on total domination—is pigweed, aka palmer amaranth. It can grow up to ten feet high in the shape of a ponderosa pine with a stalk the width of a corncob. A single plant can produce a million seeds, and a pigweed-infested field will spew hundreds of millions, raising the probability that a mutation of the plant will come along that can resist an herbicide's poison. "To a farmer, pigweed's like a staph infection resistant to every antibiotic," Heraud tells me. "Agriculture has never seen anything like it."

For decades, the chemists of Monsanto and Syngenta have struggled to make products that are molecularly "selective," meaning lethal to weeds but not to crops. The first GMO crops—Roundup Ready cotton, corn, and soy—were engineered to tolerate herbicides so that the chemicals could be sprayed indiscriminately. The solution worked until it led to the overuse of certain chemicals and, in turn, superweed resistance. In 2006, an Arkansas cotton farmer noticed that the Monsanto's Roundup he was spraying on his cotton fields wasn't killing the pigweed the way it used to. Two years later, there were 10 million acres of Roundup-resistant pigweed in the United States; by 2012, there were 30 million acres. Today, herbicide-resistant weeds infest 70 million acres of crops. Chemical companies have responded by both ratcheting up the amount of chemicals applied and reformulating old high-potency chemicals like dicamba and 2,4-D, but this approach has brought its own host of problems. Dicamba has caused chemical drift, damaging millions of acres of neighboring crops. Conflict between neighboring farmers over the use of Dicamba has become so heated that one disagreement ended in murder. Pigweed, meanwhile, has kept on dropping its tiny genetic bombs by the trillions throughout America's farmlands.

If robots can prevent herbicides from having contact with crops at all, it means that eighteen classes of government-approved pesticides previously considered too damaging to crops suddenly become viable. "We're ratcheting down the volume of chemicals that need to be used while also expanding how many types of chemicals can be used," says Heraud. In other words, Blue River could be the worst thing that could happen to the herbicide industry, or it could open the field to new products.

PST PSSST PST psssssst pst—tiny sniperlike bursts of herbicide are being shot out from 128 nozzles across eight rows of cotton plants. Patches of blue ink land on clumps of weeds in perfect rectangles, some printer-paper-size, others a thumbnail.

It's a steamy midsummer day, and we're deep in the heart of cotton country. Heraud is testing See & Spray, his first robot weeder, in cotton fields that belong to Nathan Reed, a thirty-seven-year-old third-generation farmer who cultivates cotton, corn, rice, and soy on 6500 acres in Marianna, Arkansas. Heraud is starting with cotton because those fields are planted first and they generally have the worst weed problems. Once See & Spray has earned its chops in the cotton fields, it'll move on to food crops.

Marianna looks like many small towns in the Mississippi Delta—population 4000, median income $24,000, a farming community suffering from record-low crop prices. Many of the houses at the city center, once beautiful painted-lady Victorians, are now abandoned, their porches sunken and windows broken with kudzu crawling in, evidence of the one resource this town remains rich in: weeds. This region is

See & Spray's maiden voyage

one of the weediest in the world, in fact, which makes it a good proving ground for Heraud.

See & Spray is hitched to the back of a tractor and rumbles down Reed's field at twelve miles per hour—standard tractor speed. A dome of fabric that looks like a huge white hoopskirt protrudes off the back of the tractor to protect the bot from dust and rain. Eight computers are stacked sideways underneath the dome, and above the shrouded robot three large tanks are filled with water dyed electric blue—a faux herbicide for the test run.

A software engineer is in the tractor cab looking at a ThinkPad that shows an aerial view of the ground beneath the robot—live-feed video on the screen is a composite gathered by twenty-four cameras rigged to the computers. It shows cracked brown soil with cotton seedlings poking up about three inches tall and a random assortment of weeds that to my untrained eye are indistinguishable from the cotton plants. The robot does the differentiating for us: the screen shows circles around the cotton plants, and squares—dozens of them overlapping—around the weeds.

See & Spray is scanning the plants, Heraud explains, and within thirty milliseconds—about a tenth of the time it takes for you to blink your eye—it has distinguished the cotton from the weeds, decided how much and where to spray, and moved on to the next row.

"There's a misfire—you woulda murdered my cotton plant," Reed ribs, pointing to a seedling shot with blue.

"That's why we don't use the red dye," Heraud rejoins. "It'd look too gory."

He's not kidding—in its early days, the LettuceBot did, in fact, murder entire fields of lettuce. Its nozzles sprang leaks and dripped hyperconcentrated fertilizer on acre after acre of seedlings. Heraud, known for his humility and gravitas, got on a plane and went to make things right with the affected farmers in Yuma and Salinas. His team fixed the problem by adding an automatic abort function to nozzles that drip for more than five seconds, then they thinned the farmers' next 100 acres for free.

On Reed's field, we're noticing a lot of blue-spattered cotton plants, while the weeds next to them are untouched. The machine is getting confused because some of the cotton is runty and withered—not as healthy-looking as the cotton that See & Spray's been programmed to recognize. Redden has trained the bot to identify plants in much the same way you'd train a toddler to recognize, say, a spoon, and then differentiate it from a fork. You first show her a spoon, identify it, and over time show her more spoons—oval spoons, round spoons, big and small spoons, plastic and metal spoons, bent spoons. Eventually the child discerns that the many different permutations can all be classified as spoons, and are distinct from forks and other utensils. Similarly, the bot needs to be fed first hundreds, then thousands, and eventually millions of images of cotton to learn the many variations of the plant, understanding how its leaves change shape and texture, how the plants look when they're sickly and healthy, during all stages of growth, and that all of it adds up to cotton. The robot's ability to draw from this image archive and make distinctions and decisions is called deep learning.

The Blue River team built the memory of See & Spray by going to a cotton farm in Australia, hitching a video camera to a shopping cart, and spending three months pushing it around different cotton fields, uploading about 100,000 images of cotton. But the Arkansas cotton, struggling in a wet, cold spring, isn't looking enough like the

Australian cotton for 100 percent accuracy. Each day, for a fortnight, Heraud's team will take thousands of new images of cotton, and each day the robot will become more accurate. And within a year, by mid-2018, these bots will have achieved more than 95 percent accuracy, having evolved from toddlerhood to adulthood practically overnight.

For now, though, as we amble down the field, See & Spray is making childish mistakes. Suddenly Heraud slaps his thigh. "Nailed it!" he shouts, breaking his characteristic composure. He's looking down at a cotton plant engulfed around its perimeter by a nasty weed. The machine has sprayed a ring of blue liquid just on the weed and spared the struggling seedling at the center. With an index finger, Heraud tousles the leaves of the seedling. "Say it's a baby corn or soy plant—this is what it looks like to keep chemicals out of the food system." It strikes me then that Heraud's invention is as nostalgic as it is futuristic—that its purpose is to redress the problems that have been wrought over many decades by less intelligent technology.

Where Heraud sees lowered chemical use, Nathan Reed sees savings. Herbicides comprise more than 40 percent of Reed's operating costs—over half a million dollars a year. On a single acre of cotton, he typically uses about 25 gallons of an herbicide solution made with Monsanto's Roundup; after two weeks of trails on his fields, the See & Spray

The Blue River bot sprays the weed and spares the seedling

robot could manage his weeds with less than 2 gallons of the herbicide per acre. A robotic weeder also means he doesn't have to buy GMO seeds that have been engineered for resistance, which cuts his seed costs by about three-quarters. But Reed, like most farmers, is struggling to get by, and robotic weeders will be an option for farmers in Arkansas—or, for that matter, on the growing industrial farms of Kenya—only if Heraud can sell his invention at a competitive price.

IN RECENT DECADES, the world has lost a third of its arable soil because of erosion caused by mechanical tillage and damage from industrial chemicals. More than 1 billion pounds of pesticides are used in the United States each year—about a fifth of the 5.6 billion pounds or so that are applied worldwide.

The use of herbicides on American farms began in the 1940s with the application of a toxin developed by chemists in World War II known as 2,4 dichlorophenoxyacetic acid, or 2,4-D, to lawns and grain fields, but it exploded twenty years later with the discovery of glyphosate in the late 1960s. John Franz, a young ace developing flame retardants for Monsanto, was brought into the company's ag division to help develop a nontoxic herbicide. 2,4-D had been the product of biological weapons research, and Monsanto wanted to find less dangerous ways to kill weeds. Franz discovered that glyphosate inhibited a key growth enzyme found primarily in plants and had no apparent effect on mammals, birds, fish, or insects. Monsanto released the chemical under the brand name Roundup, billing it as the safest herbicide in history. This was actually true. "Glyphosate is probably the best and most benign herbicide ever invented, but anything used in such huge volumes is bound to backfire," says Adam Davis, a chemist who works with the Environmental Protection Agency and the USDA. "As the medical adage goes, the dose makes the poison."

In the two decades between 1996 and 2016, the use of glyphosate worldwide shot up more than fifteenfold. During that same time, the percentage of Americans who tested positive for glyphosate (based on urine samples analyzed by the EPA and the National Institutes of Health) increased by 500 percent. Today glyphosate is still used on nearly all—more than 95 percent—of U.S. crops, even as the superweed populations grow and as evidence of human health consequences mounts. A subsidiary of the World Health Organization stated in 2015 that glyphosate in high levels is a "probable carcinogen"—after four decades of rampant use in the United States. Recent studies have also tied high levels of glyphosate and other government-approved pesticides not just to cancers, but to allergies, ADHD, and Alzheimer's.

There's also growing evidence that popular herbicides can harm soil microbiology, in particular the activity of earthworm populations that naturally aerate and fertilize soil. Chemical fertilizers, which are applied to soil in far greater volumes than herbicides, also pose a problem. They offer a good short-term nitrogen boost but over time the excess nitrogen can overstimulate soil microbes and cause them to self-destruct.

For all the potential impacts of chemicals on soil, says Nathan Reed, a bigger challenge for soil health, both near- and long-term, is their seemingly benign alternative—mechanical tillage. The churning of soil by tractors to kill and suffocate weeds is the method used by most conventional farmers and nearly all large-scale organic farmers, yet it causes erosion. Currently, U.S. soils are degrading ten times faster than they can be replenished. Tilling also dries out soil—it was a key factor causing the Dust Bowl crisis in the 1930s—and disturbs the microbiome. "If you're a microbe or an earthworm and you've had an all-you-can-eat buffet for six months provided by the healthy root systems of your crops, and suddenly they're churned up and gone, are you going to stick around?" says Reed. "No, you're not."

Reed practices "no-till" farming, in which the farmer forgoes tilling altogether, allowing crop residues to decompose naturally into a kind of fertile carpet on top of the soil. "You plant your new seeds right into that crop waste," says Reed. "It's not pretty, but it works." He plants a cereal rye after harvesting a crop, which grows up for several months and replenishes soil nitrogen naturally before he mats it down and plants his cash crop. The single biggest advantage of the See & Spray, as Reed sees it, is the role it can play in easing and expanding no-till farming. He gets about 15 percent better productivity—200 pounds more cotton and about 30 bushels of corn per acre—on his no-till fields than on those he ploughs. No-till keeps soil moist and conserves irrigation costs. Nathan's cover crops have cut his irrigation costs in half, by about twenty-five dollars an acre.

No-till farming also locks carbon in the ground. As crops grow and photosynthesize, they sponge carbon dioxide from the atmosphere. While some of the plant is harvested, the rest—the plant roots and residue—stays in the field, decomposes, and becomes soil. Ploughing releases the carbon back into the atmosphere, while no-till keeps it sequestered. "If all the world's farmed land was no-till, it would go a long way to solving climate change," says the USDA's Jerry Hatfield.

The major downside of no-till is weeds, especially pigweed (the seeds like to stay on top of soil and will die only if they're ploughed under). "Without mechanical tillage, most farmers resort to heavier use of chemical herbicides," Heraud explains, which is to say, no-till is often chemically intensive. It was the onset of herbicides that first enabled the no-till practice to spread in the 1970s and '80s, in fact. "Organic farms today can't easily do no-till because they don't use herbicide," says Heraud.

For now, only about a fifth of U.S. cropland is farmed full-time using no-till—and less than 10 percent of cropland globally. This concerns Hatfield: "Let's call a spade a spade—there's been a pretty minimal increase in no-till in recent years. Farmers are locked into old habits, and those habits will have to change if they want to survive the climate pressures ahead." In theory at least, technology like See & Spray can make no-till easier and more affordable. Were this practice to become the dominant mode of farming among organic and conventional farmers alike, it could mean, says Hatfield, "a revolution in soil health and carbon sequestration."

But such a shift will also require a better understanding of soil itself. Circa 1510, Leonardo da Vinci said, "We know more about the movement of celestial bodies than about the soil underfoot." This hasn't changed much. It's also true that a single teaspoon of healthy soil contains billions of microbes—more than there are people on this earth. Within that spoonful there are some 10,000 to 50,000 different species, including vast populations of nematodes (tiny worms), microarthropods (microscopic bugs), and single-cell protozoa. An article in *The Atlantic* titled "Healthy Soil Microbes, Healthy People" explains that the diverse bacteria and fungi in soil "serve as the 'stomachs' of plants. They form symbiotic relationships with plant roots and 'digest' nutrients, providing nitrogen, phosphorus, and many other nutrients in a form that plant cells can assimilate." All this soil science fascinates Heraud, who holds that "the best thing we can do for our future food supply is to protect the health of the soil."

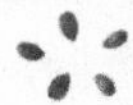

AFTER A LONG day in the Arkansas cotton fields, we settle into some Barcaloungers in the living room of a duck-hunting lodge Heraud has rented to house his team during their visit. The scene feels like the cast of HBO's *Silicon Valley* has stumbled onto the set of *Deliverance*. Twelve techies living frat-house-style in a hunting lodge loaded with taxidermy, two pinball machines, and crates of empty beer bottles. Here, Heraud tells me his plan for affordably producing his machines at mass scale. It hinges on the agricultural colossus, John Deere, which is about to announce its plan to acquire his six-year-old start-up for hundreds of millions of dollars.

Heraud isn't concerned about giving up his independence. "Deere gives us far bigger chances of having a worldwide impact," he says. "Without this parternship, we're a tiny company that might succeed or fail. A rookie with two weeding prototypes. Any more scary scenarios"—like the leaky nozzles on the LettuceBots—"could be a fatal blow."

The acquisition also means Blue River can release its first See & Spray by 2020—several years sooner and on a much larger scale than it could have without Deere's army of mechanical engineers, forge factories, and ten thousand dealers around the world. John Teeple, the director of advanced technology at Deere who courted Jorge and his team for months before the deal went forward, tells me, "You might think, Oh, wait a minute, what's this conservative 180-year-old company doing with a Silicon Valley start-up? Well, we're helping define a new era of agriculture." He continues, "What's happened in the last two or three years in farming with data and digital technology is a more immense change than we've seen in decades. The industry is transforming before our eyes, and it was clear to us that Jorge and his team were way out on the front end of that change."

Heraud's next step, with Deere's backing, will be moving Blue River's bots beyond herbicides to fertilizers. Farmers typically spend up to ten times more annually on fertilizers than weed killers—about $150 billion a year. But the shift is a big leap for a robot. The bot must gather a range of visual signals—the colors, sizes, and textures of a plant's leaves—and from this data extrapolate how healthy the plant is and how much more nourishment it needs. "It's a ton more processing power, but it's doable," says Heraud.

The next link in this technological chain could be a kind of agricultural Swiss army knife: a robot that can treat each plant individually, not just with herbicides but also with customized fertilizers, insecticides, fungicides, and irrigation, all at once, delivering only as needed. The implications of plant-by-plant—rather than field-by-field—farming is not just vast reductions in chemical usage. It could, at least in theory, end monocropping—the cornfields and soy fields as far as the eye can see that have become the new normal. Monocrops leach soil nutrients and put our food supplies at risk, since single-crop fields are more susceptible to blight and catastrophe.

Heraud argues we've been segregating crops in part because our equipment couldn't handle more complexity. Robots that can tend plants individually could also support the traditional practice known as intercropping—planting corn in with complementary crops such as soybeans and other legumes. In this way, Heraud's robots would help restore the sustainable practices of diverse, smaller-scale farms and help redress the problems created by the Green Revolution.

Danielle Nierenberg, president of Food Tank, a think tank that advocates for sustainable agriculture, isn't so sure. "I'm not particularly comforted by this vision of AI farming," she says. "Lots of questions need to be asked, like which chemicals will be deployed by these bots? And what farming jobs will be lost as robots take over? And which of the many problems inherent with industrial agriculture will persist even as we cut back on herbicides?"

One of those problems is the potentially coercive power of monopolies. John Deere is cast as a villain in the right-to-repair movement, in which urban and rural DIY-ers are fighting to pass laws against the use of proprietary software and hardware that makes it nearly impossible for individuals to fix their own gadgets or machines—a problem that maddens an owner of a $500 iPhone but could economically devastate an owner of an AI-enabled $200,000 tractor.

For Nierenberg, it's easy to feel hopeless about this agricultural future, one in which a farmer's success hinges on fealty to a few monolithic corporations and their patented robotic offspring. It seems to have the makings of a perfect techno-dystopia. Deere's ability to make farmers dependent on the usage and, increasingly, the maintenance of its specialized equipment bears relation to Monsanto's system of locking farmers into its herbicides and seeds. There's also the possibility, however remote, that a software-dependent food system will become vulnerable to hackers who could manipulate the dosages of toxic chemicals on the fields.

Heraud prefers not to fixate on worst-case sci-fi scenarios. "It's not either-or—should we do technology or agro-ecology, sustainable farming or industrial farming?" he says. "It's *both-and*. We need all solutions." He brings me back to the connection he made as a kid between farms and factories. "A hundred years ago, factories were a nightmare, spewing black smoke, with terrible working conditions and people dying. A lot of agribusiness is in that state right now, with massive inefficiencies, harmful chemicals, a huge carbon impact. But compare that with modern factories, designed to be smart, automated, safe for the environment and humans, ergonomics on every single job—they've turned around." The happy paradox, Heraud insists, is that "robots don't have to remove us from nature—they can help us restore it."

It's a hopeful articulation of a third way approach to agriculture. But even if you rule out the hidden or unintended environmental consequences of this technology, robots can only be one small part of a solution to future food production. There's a limit to what "seeing" robots like Heraud's can do. For one thing, Heraud's bots are confined to analyzing problems above the surface of the soil, when in fact so many problems in industrial agriculture are below it. Here too, Heraud stresses the potential of digital tools: electronic soil sensors are now beginning to analyze aspects of soil health down to minute microbial activities, and they can wirelessly deliver that information to farmers on the fly. Heraud and his team have also developed drones with infrared sensors that can cruise above crops to monitor how well the plants are absorbing and reflecting light, assessing their growth and health.

These drones, sensors, and robots are part of a growing network of data-collection and -crunching devices that can feed increasingly detailed information to farmers about their crops. The broad term for this digital farming universe is "precision agriculture," and I first get the chance to explore it inside, of all places, a glass-and-steel office build-

Farmer drone

ing in downtown Shanghai. There, I meet urban software engineers and data analysts employed by one of the largest organic farming enterprises in China who are managing the growth of fruits and vegetables by toggling the buttons and swiping the screens of their iPads. It's an enterprise that puts a twist on the farms-as-factories analogy Heraud dreamed up as a kid—this is more like *Ready Player One.*

CHAPTER 5

Sensor Sensibility

Technology is a useful servant but a dangerous master.
—CHRISTIAN LOUS LANGE

IT'S ALWAYS RISKY to invest money, to say nothing of hope, in new technologies. This is especially true in the realms of food and agriculture, where over many centuries and especially in recent decades, there have been countless attempts at tech-driven solutions that lacked humanitarian purpose and even utilitarian function. One notorious example of this is Juicero, the Silicon Valley start-up that raised $120 million in funding to produce a pricey, internet-connected machine that made personal cold-press juices. It didn't take long before customers figured out that the juice packets needed neither a machine nor the internet to be pressed (they could be squeezed by hand) and the company crumbled. "To say the future of food will be high tech tells us little about the values of the food system we're building for future generations," warns an article in the progressive magazine *The Nation* titled "The Future of Food."

No doubt we should be wary of ideas in agriculture that smack of overengineering—and the success of any technology will depend on how it's applied. New genome-editing methods like CRISPR could be used to help engineer more nutritious and resilent crops—or to worsen the existing problems of industrial farming. AI robots could be designed to support more diversified food systems and to advance agroecology, or to undermine it. "Tech should serve humanity, not the other way around," said Tim Cook, CEO of Apple. And yet, it often doesn't.

Perhaps nowhere does the intersection of technology and agriculture present greater risks and more significant potential rewards than in China, the world's biggest food producer. China has risen to the status of global superpower faster than any nation in history and generates one of the largest global GDPs. Despite this—or because of it—China's quest to feed its 1.4 billion citizens has become an almost inconceivable challenge. The rocketing demand for food from the country's urban middle class is increasingly at odds with its limited arable land, severe freshwater shortages, and choking smog. "If you want to know what it'll look like to grow food in a more populous, environmentally stressed, postindustrial world, visit the farms of China," said Manuela Zoninsein, a Beijing-based entrepreneur and one of several friends who encouraged me to visit. "It's actually amazing that China's food system feeds as many people as it does," said Brian Heimberg, a Californian who moved to Shanghai to join a clean-tech venture capital firm. "You have to see it for yourself."

I would do so after reading an article in *Caixin*, a Chinese business publication, about Tony Zhang, an entrepreneur trying to cultivate thousands of acres of organic farmland in and around Beijing and Shanghai. When we talked before my visit, Zhang told me of his plans to build "the Whole Foods of China" and promised to show me "the leanest, smartest, most automated organic farming system there is." It was an intriguing, if overzealous, pitch that drew me into a more complex and sobering story about Chinese food production, one with strange details—from cabbages doused with formaldehyde to military-led soil cleanup efforts—and unexpected plot twists, including the mysterious disappearance of Tony Zhang himself.

THE LANDMASS OF China is about the same size as the United States, but its farms sustain triple the population. The western half of the country is covered in mountains. The northeast is cold and dry. The southeast is temperate and historically very fertile. This is the part of the country that's getting inundated with urban development. In the last two decades, China has lost crucial farmland to its spreading cities. All told, the country has 0.2 acres of arable farmland per citizen (America has about five times that amount), and much of that acreage has been poisoned with pollutants.

While the United States relies on rural regions for its food production, China's farms are mostly located in the greenbelts that surround major cities. For all its engineering expertise, China still has a nascent national road system and limited storage facilities, which makes food transport over long distances difficult and costly. Beijing and Shanghai get more than half of their food from local farms. This would seem like an environmental advantage, given the lower fuel costs and carbon impact of local food systems, but the urban areas of China are also inundated with contaminants. In 2015, the mayor of Beijing deemed his city of 24 million people "unlivable" due to smog; the air quality has since improved, but pollutants are still many times more concentrated than the WHO's maximum safe level. These toxins rain down on the land and into aquifers during storms. And for decades, poorly regulated industries have been dumping chemicals into China's waterways. Now a quarter of the country's lakes and rivers have been deemed unfit for human use, and three-quarters of the watersheds that supply China's thirty fastest-growing cities have "medium to high" pollution levels.

In 2014, the Chinese government published a national soil survey which showed that about 20 percent of the country's farmland (an area about the size of the Netherlands) is contaminated by chemicals and heavy metals. Poisoned soil taints the food that's grown within it—a serious public health concern—but it's also less productive than healthy soils, which means farmers with contaminated plots must add even more chemicals to boost production. China's farmers are using, on average, up to four times the levels of fertilizers and pesticides used in the United States. Like farmers in Kenya and elsewhere, many Chinese growers have adopted Western agricultural practices with high chemical inputs, but more aggressively and on a much bigger scale. The meat producers of China, who've also been facing surging demand, are running increasingly concentrated livestock operations, and some have resorted to counterfeit products.

The pressure on all of China's food producers to ramp up yields has given way to crimes so severe that the government created a special police unit to rein in corruption. In recent years, food inspectors have reported arsenic in apple juice, melamine in milk, rice laced with cadmium, diseased pigs sold as fresh pork, rat meat sold as lamb, and watermelons exploding from chemical growth accelerators. In the last decade, the authorities have executed two workers for food-related crimes and imprisoned thousands more. China also has a rampant problem with food spoilage. "Two hundred million Chinese every year get sick from bacterial contamination of food," says Junshi Chen, senior adviser to the China National Center for Food Safety Risk Assessment.

There are efforts to both curb and capitalize on these exigencies. "Chinese consumers have grown wary about food safety—that's one of many factors that have made this a potentially huge market for imported organic foods," Eric Newman, director of sales for the Wisconsin-based dairy cooperative Organic Valley, tells me. Chinese investors have ploughed billions of dollars into overseas food companies, including the world's largest pork producer, Virginia-based Smithfield Foods, and Australia's largest dairy.

But many have been trying to improve China's food system from within. There are an estimated 200 million farms in China; the vast majority span small plots of land and employ techniques that date back centuries. Tony Zhang was one of the first among a new breed of food growers in China who've begun to rethink sustainable food production on a large scale—some with significant investments in high-tech irrigation systems, soil sensors, modern seeds, robotics, and data science.

When I first visited Tony's Farm, as Zhang named his enterprise, in 2014, it encompassed more than ten thousand acres in eight provinces.

Tony Zhang

He was reaching about 200,000 customers and growing 120 different types of organic fruits and vegetables. Zhang's goal was to blend nostalgia with futurism, to merge traditional agricultural values with new tools and techniques that could help cure a food system in crisis.

TONY AND I FIRST met on a terrace outside his downtown Shanghai office. Immaculate in a starched olive blazer, white T-shirt, and tailored jeans, Zhang was sipping a glass of honeydew juice the color of kryptonite. He opened our conversation by gushing about the quality of his melons. "They are the best you have ever tasted," he said through a translator. Zhang, a fifty-one-year-old native of Sichuan, had at this point reeled in $40 million in funding from Western and Chinese investors, along with millions more in government subsidies for his land cleanup. He was appealing to the country's fast-growing population of elite urbanites who are driving up the demand for organic products.

Tony had a flair for extravagance—his silver Bentley was parked not five yards from our table on the pedestrian walkway—but he also had legitimate standing in China's foodie world. Before founding Tony's Farm, he was the owner of Tony's Spicy Kitchen, a popular restaurant chain he expanded to thirty-three locations in six years. Zhang had a creaseless face and a preternaturally calm demeanor, yet was referred to by some colleagues as "Spicy Tony." During our travels, his assistant would offer him a dish of ground-up Tien Tsin—the extra-hot red Chinese peppers used in Kung Pao chicken and other Sichuan and Hunan dishes—to sprinkle onto his food and even stir into his frequent cups of black coffee.

Zhang's plan was to scale his farming enterprise as he did his restaurant chain. "I admire the model of Earthbound Farm," he told me, referring to the California-based organic produce company founded in 1986, which now grows about 50,000 acres of fruits and vegetables on three continents. "It took twenty-eight years to grow the company [to this size]. I am confident it will take me ten years to become as big as Earthbound."

In retrospect, this should have been a red flag; growing too fast is a downfall of start-ups in every industry—especially in agriculture, and especially in China. "Organic farming in China is extremely time-intensive; it takes a lot of time to prepare the soil and requires gradual growth," explains Shi Yan, the young founder of the Beijing-based organic farming cooperative Shared Harvest. "That's something large investors don't have time for." An added complexity for Zhang was that his model was unproven—not the same as Earthbound's, exactly, nor Whole Foods', but a hybrid. Tony was positioning himself as both a farmer and retailer. Straddling the two categories could be an advantage, but also a cost. As a farmer, Zhang had to accept, for instance, the dizzying price of cleaning polluted soil and water. As a retailer, he faced the enormous expense of building and maintaining a logistics system for storing and transporting his produce.

Tony's Farm was designed to function like a giant online vegetable box scheme. At its height it had thousands of subscribers—families, companies, schools, restaurants, and markets in Shanghai and Beijing. They would submit their orders online, and the food would be delivered on demand, within twenty-four hours. Most subscribers received weekly deliveries of staples like carrots, tomatoes, peppers, and berries,

A packaging floor at Tony's Farm

along with Chinese delicacies like red amarynth leaves, sponge gourds, purple begonia greens, wood ear mushrooms, and yard-long beans. Subscribers could also order organic meats, eggs, oils, grains, and pantry products that Zhang sourced from other suppliers.

He maintained a farm-to-kitchen logistics system to store and transport produce from fields to customers' homes, which included managing a costly fleet of dozens of refrigerated trucks. Zhang passed some of these costs on to consumers, with prices about three times the average for conventional Chinese produce (the equivalent of about $2.65 per pound vs. $0.80). Earthbound and Whole Foods customers, by comparison, pay a smaller premium in the United States, where conventional produce is not as cheap. But Zhang said he believed his consumers would shell out for peace of mind: "Trust and quality are the DNA of our brand."

TONY'S FARM GREW out of, more than anything else, nostalgia—"a simple desire for the flavors of my childhood," he says. The way

I often pine for my mom's four-cheese lasagna, Tony spent much of his adult life longing for his grandmother's Sichuan beans and pork, made from the green beans, garlic, peppers, and pigs raised on his family's farm. Zhang grew up in a small agricultural village outside of Yibin, a city in Sichuan province. It was not far from the kind of setting you see on postcards: emerald mountains, bamboo forests, waterfalls, and panda bears. Zhang was an only child, and he spent whatever free time he had working the land or wandering the natural world. It didn't occur to him then that *nongimin*, the Chinese word for farmer, is also a derogatory word that means "peasant" or "low-class." He says that growing up he had "a special feeling about agriculture—it was a privilege."

Zhang excelled in high school and enrolled in the Sichuan Agricultural University at the age of sixteen, graduating at nineteen. He worked in Yibin's department of agriculture, and moved to Shanghai at twenty-eight to work for a foreign trade company selling products from pharmaceuticals to metals. Six years later, after rising through the ranks, Zhang bought the company—a move that helped him fund his future enterprises. He'd become so homesick for spicy Sichuan cuisine that in 1997 he founded Tony's Spicy Kitchen.

As Zhang's restaurant enterprise grew, he worried about the quality of his ingredients. "The taste of the produce wasn't right," he remembers. "Too bland. Bad texture." Zhang found out which farms his vegetables were coming from and went himself to investigate. He noticed that the farmers kept two separate plots of land—one organic farm for their family, and a separate commercial farm treated with chemicals to speed crop growth. He was outraged that his suppliers didn't consider their commercial produce safe enough for their own tables.

Zhang deepened his research. Around this time, stories were surfacing in the media of farmers dousing lettuces and cabbages with formaldehyde (a chemical that can be lethal if ingested) to keep the produce from rotting during storage and transport. Scientists had begun studying soil quality on farms in regions such as Hunan province, where many rice producers operate farms near mines and smelter plants releasing large amounts of cadmium, a heavy metal toxin linked to neurological damage and cancer. One report from China's Ministry of Agriculture revealed that more than a quarter of the rice the government had sampled nationwide had excess lead, and a tenth had excess cadmium. Soil contamination, Zhang realized, was one of the largest and most neglected problems in the country.

In 2005, Zhang sold his restaurant chain, leased his first 290 acres of land in Nanhui, a suburban district of Shanghai, and incorporated Shanghai Tony Agriculture Development (later changing the name to Tony's Farm). Zhang then began his first arduous effort to clean up contaminated soil, and four years later, in 2009, he began to deliver organic vegetables to his first batch of customers. Within months, he had thousands of subscribers.

In the early days of his enterprise, Zhang reconnected with Jiang Hong, a professor he'd had at the Sichuan Agricultural University and the man he later appointed chief scientist of Tony's Farm. Hong, a specialist in agricultural technology, advised Zhang that a combination of intelligent sensors and software could substantially improve the efficiency of his operation. "The costs of organic farming are so high that you must do whatever is necessary to produce the very highest yields per acre of land using the very least inputs," Hong told me. Hong brokered research relationships between Tony's Farm and Chinese universities, including Jiaotong (China's MIT). They adopted new software and data-networking tools and began to develop a smart-farming system capable of governing crop growth as well as the farm's packaging, storage, online sales, deliveries, and quality control.

When I visited the Nanhui farm, Hong showed me a field of amaranth, a purple-hued Chinese spinach, that had dozens of sensors stuck like miniature flagpoles in the soil every few rows. The sensors were long, skinny metal rods with transmitters at the top for wirelessly delivering the collected data. Hong and his team were experimenting with soil sensors throughout the farm to monitor the microclimate of each crop, gathering data on moisture and temperature, humidity, acidity, and light absorption. Like traders watching stock tickers, five technicians on Hong's team spent their days in downtown Shanghai parsing data streamed from the sensors in real time from various farm locations. If sensors detected low levels of moisture in the soil of, say, a certain potato crop, said Hong, the system would automatically activate the sprinklers that irrigate that crop, restoring only as much soil moisture as the plants needed. Zhang said the moisture sensors were helping to cut his water demands by about half, and to reduce the energy costs of pumping water to his fields.

This kind of precision irrigation can be a key advantage given China's drought problems and the high cost of freshwater. Contamination has only compounded the problem of China's inherent freshwater shortages. In Beijing, which is in an extremely dry region, residents are allotted just 100 cubic meters of water per year—compared with the 2840 cubic meters U.S. citizens consume on average. Similarly, water is rationed to many of China's farmers, and any usage over the quota can be very costly.

Hong was designing a precision-farming system to monitor and meet the specific needs of dozens of crops. With commands via iPad from Hong's technicians in Shanghai, the system could automatically deploy certain inputs from many miles away: here a nutrient boost; there an organic pesticide. What Zhang envisioned was a form of remote-control farming. But crops would often have problems that required hands-on care. In these cases, the Shanghai office could dispatch field hands to troubleshoot. If the temperature sensors noticed a sudden increase in soil heat or microbial activity, this might indicate the presence of a harmful bacterium or disease at the plant roots; if humidity sensors shot up, the crop in that area might be vulnerable to fungus.

Zhang described the technology as a means to cut back not just on water and chemical use, but also on labor costs. The entire farming staff of Tony's Farm in its heyday was about two hundred people—a fifth of what it would have been without the precision-farming tools, said

Zhang: "Organic farming is much more labor intensive than conventional agriculture, so this gives us an edge."

Zhang also wanted his customers to be able to participate remotely, as observers, in the farming of his food. He asked Hong to develop an app with a product-tracking technology that could enable customers to scan labels on the item they purchased with their smartphones to learn how it was grown, the soil and water quality of the field it was grown in, and even watch video feed of their growing fruit or vegetable (Hong set up surveillance cameras in some fields and greenhouses tracking crop growth 24/7). This would help build more trust, Zhang reasoned, with his safety-wary customers.

But while precision-farming and surveillance tools can go a long way toward improving efficiency and building trust with customers, they also add huge expense. Shi Yan of Shared Harvest has kept technology use on her organic farms to a minimum for this reason: "Modern technology solves some problems but creates others. You must hire skilled people to manage the tools and software. Most organic farmers are working on a small or medium scale and cannot afford that. The costs can be crippling."

AN EQUALLY ONEROUS cost is soil cleanup. "When I started in the organic industry, people thought I was crazy," Zhang told me. "Land and water cleanup come at great expense." Zhang said those costs amounted to tens of millions of dollars in the first decade of his enterprise. The Ministries of Agriculture and Technology have invested trillions of yuan in subsidies to help farmers speed the cleanup process. Zhang benefited handsomely from this, but still he had to spend a large share of his own funding on soil cleanup.

Zhang entered the organic farming industry when there was still little public knowledge about the full magnitude of the soil contamination crisis in China. It wasn't until 2014 that the Chinese government

Farmers weeding in Xinjiang Province

disclosed that fully a fifth of the country's farmland was contaminated with heavy metals—information that had been guarded as a secret before it was leaked. In 2016, chemical residues became a national story when students at Changzhou Foreign Language School got sick after complaining of a terrible smell. Some were diagnosed with lymphatic cancer. Investigators found that the campus had been built next to a defunct chemical dump. The local government had bought the land and hired engineers to place a thick layer of clay over the chemical waste, but toxins had seeped through the sealant.

Around that time, a report was published in *Environment International* by a team led by Yonglong Lu of the Research Centre for Eco-Environmental Sciences in Beijing, showing a strong association between industrial contaminants on farms and the prevalence in nearby populations of diseases, including hepatitis A, typhoid, and certain cancers. By the end of 2016, the Chinese government had issued a plan to clean up 90 percent of contaminated farmland by 2020. In 2018, it deployed a regiment of sixty thousand soldiers to plant forests of land-restoring trees across an area roughly the size of Ireland.

Yonglong Lu tells me that while the Chinese government "is putting great effort into cleaning up the soil," the processes for cleanup are laborious, time-intensive, and very costly. Soil remediation methods include excavating polluted soil and replacing it with clean soil, and, more commonly, using chemical and bacterial agents to degrade the pollutants. "These processes are not always effective," says Lu.

Even phytoremediation, the traditional method of planting flowers and weeds (sunflowers and ragweed, for example) and trees (willows and poplar) to absorb the heavy metals in the soil, is problematic. When the plants are removed, stems and roots break off and the contaminants often end up back in the soil. Lu maintains that "phytoremediation is very hard to do effectively on a large scale." That may be why some Chinese plant scientists are trying to engineer seeds for crops that won't absorb toxins through their roots—a solution that may hold promise, but flouts the principles of organic farming.

According to Qiu Qiwen, head of the soil environment department of the Ministry of Environmental Protection, pollution cleanup can cost about $18,000 per acre of soil, an extraordinary sum when you consider that China has about 200,000 acres of heavily contaminated farmland. To make matters worse, Chinese soil is also inundated with fertilizers and pesticides. *The Economist* reported in 2017 that pesticide use has more than doubled in China since 1991, and "the country now uses roughly twice as much per acre as the worldwide average. Fertilizer use has almost doubled, too." While the government has tried to curb the use of agricultural chemicals, it's all but impossible to impose standard practices in a food system that encompasses hundreds of millions of individual farmers. Compared to the United States, China has more than fifty times the number of individually owned farms.

It stands to reason that the organic certification process in China has been, at best, inconsistent. "It's common in China for produce to be falsely advertised as organic," says Jay Wang, a former government employee who worked on organic certification. "Many farmers have used organic practices on only a small portion of their farm, with chemicals on the rest, and still obtained organic certification." The difficulty of enforcing safety standards in a system this decentralized, said Zhang, further promotes confusion and distrust among Chinese consumers.

DURING MY VISIT to Shanghai, I traveled with Zhang to the packaging floor of his flagship Nanhui farm, forty miles west of the city. It looked like a set from *Barbarella*—all Plexiglas and linoleum and stainless steel. Workers wearing coveralls, gloves, and hairnets were spraying piles of fresh-picked produce with water from coiled steel hoses that hung from the ceiling. They then hand-dried the fruits and vegetables and placed them in cellophane bags that bore the motto "Organic Starts from Tony's Farm." Across the room, in a huge glassed-in laboratory, goggled scientists tested samples of each newly harvested crop for bacteria and chemical residues.

Zhang strode onto the floor to greet Abby Ding, his director of packaging and distribution. As the two were chatting, Zhang snatched a bag of carrots rolling by on a conveyor belt. Inside the cellophane were eleven conical carrots and one slightly gnarled one. Zhang frowned at the rogue carrot the way a parent might scowl at an impossible child. He removed it from the bag and chucked it in the reject pile.

"But that's the way a carrot naturally grows!" countered Ding.

Zhang told her that carrots must grow straight if you want organic food to succeed in mainstream China. He recalled another scenario in which he received a complaint from a customer who found a snail on a head of lettuce. "Of course, the snail is natural, a sign of healthy soil," said Zhang. "But our customer does not want it there."

In the early days of his farming business, Zhang wanted to grow the company slowly. It troubled him to discover that "farmers don't want to spend the time necessary to grow animals and crops naturally," he said. "A pig traditionally needs to grow for two years before it's ready to eat; now it's just a few months, and then it goes to the restaurant." Zhang said he wanted to restore not just the flavors of his childhood but traditional home-grown values to agriculture: "I wanted to follow nature." He hadn't anticipated that "following nature" would eventually involve costly data networks, expensive soil cleanup, and inefficiencies like chucking geometrically imperfect carrots.

These costs may ultimately be what pushed Zhang to leave the business. Now, several years after my visit, it's unclear if Tony's Farm is producing carrots—or any produce—at all. In 2018, I got word from an investor in Tony's Farm that Tony himself had quietly left the company and sold the majority of it to a real estate firm and an insurance company. "The costs of production were extremely high, the logistics incredibly complex," reasoned the investor. Tony's Farm still presents itself as an active company, on its website, at least, but everyone I spoke to in the organic food industry told me it was a facade. My friends in Shanghai reported that they no longer saw the once-ubiquitous Tony's Farm delivery trucks driving through their neighborhoods. "This was a sustainable-food company that wasn't in the end sustainable," the investor tells me.

For months, I tried to reach Tony to get his version of the story but couldn't track him down. There was no answer at any of the corporate headquarters; everyone I'd met at the company appeared to have quit. I eventually came to realize that Tony's Farm has produced, if not abundant and sustainable food, then a cautionary tale and perhaps even a precedent-setting vision. It's one of countless examples of start-ups that have tried to grow too big, too fast, while shouldering the enormous costs of technologies that haven't yet matured.

Findle Zhao, who manages sales for Organic Valley in Asia, describes Tony's Farm as an idea that was ahead of its time. "It's an important story to tell. There will always be the examples like this—the early adopters. Even if they don't succeed, they make a path for others who will use those tools in the future, when costs have come down."

Long before he sold off his company, Zhang had described his enterprise to me, at least in part, as a form of public health advocacy and environmental do-goodism. He understood that with products three times the price of conventional produce, he was trying to sell Tesla-caliber sustainable food. Still, he insisted his products should be seen as necessities for the emerging middle class, not luxuries for the wealthy elite. He described the biggest barrier to his success not as the cost of technology and soil cleanup, but instead as consumer education. He insisted that for his company to succeed—and lead the way toward sustainable agriculture—it wasn't so much his prices that needed to change but the mind-set of Chinese consumers: "The problem is that conventional food in China is so cheap—a fraction of what it costs in the U.S." These prices don't reflect the hidden costs of eating contaminated food, which can be very high. It's better for middle-class Chinese to invest in safe food that costs more, Zhang argued, than to buy cheap food that makes them sick. "The consumer mentality should be: Let's pay the farmer, not the hospital."

Junshi Chen, at the China National Center for Food Safety Risk Assessment, cautioned that luxury legumes, even with economies of scale, are a far cry from a public health panacea. "We cannot depend on organic farming to solve the food-safety issue for 1.3 billion people," he said. Zhang countered that his tech-focused approach to organic farming would eventually trickle down into mainstream farms, and "that benefits everyone."

On this last point, Spicy Tony may have been right. Since he founded Tony's Farm in 2005, the use of soil sensors, smart data networks, and other precision-farming tools have proliferated worldwide, including on large conventional farms in China, well beyond the organic sector. And while most rural small-scale farmers don't have the knowhow, internet connectivity, or wealth to invest in such tools, that may change in the

Tending seedlings in a Tony's Farm greenhouse

coming decades. Xiaoming Zhang, an organic farmer who cultivates 500 acres in and around Beijing, tells me his farm is "fully equipped with digital instruments, cameras, and onsite technical personnel." Xiaoming adds that he named his farm Noah's Organics after Noah's Ark "to convey a message of safety in a time of concern."

WHATEVER THE REASONS that Tony Zhang abandoned his farming enterprise, it was at a time when the broader industry was on the rise. Sales of organic foods in China surpassed $7 billion in 2018. Organic certifications more than doubled between 2013 and 2018, according to government data, even as the certification process has become more costly and stringent.

The online organic retailer FieldsChina.com now delivers products to two hundred Chinese cities; KateandKimi.com, which farms and delivers its organic produce, is popular among millennials. Walmart has been ramping up its organic offerings in China and has tripled its spending on food safety there. The U.S. supermarket chain Kroger recently announced that it would begin selling its organic foods line, Simple Truth, in China—not via brick-and-mortar stores, but Tony Zhang–style, through e-commerce sales.

Perhaps the biggest challenge for Zhang was pressure from his investors to stray from his core values and vision. "We want to see faster growth," one of Zhang's largest investors told me during the company's heyday, "but this is such a capital-intensive and time-consuming business." Zhang's investors pushed him to start selling organic meats, eggs, oils, and other products from outside suppliers—a move he resisted at first. They then convinced him to include off-season produce and products labeled "natural" and not certified organic. Zhang capitulated to many of the demands. He also began to move away from the urban areas and lease land in more remote parts of China, including a new 450-acre farm in Sichuan province near his hometown. Airfreighting this rural produce to cities was actually faster and cheaper than cleaning up contaminated urban soil, but the delivery and storage logistics were still wildly expensive.

Some of Zhang's investors were also drawn to another form of farming altogether—one that eliminates soil cleanup because there's no soil to speak of. Known as "vertical farms" in the United States and "plant factories" in China and Japan, these new indoor growing chambers produce crops without soil or sunlight or seasons. The plants are grown 24/7 in massive warehouses under high-intensity grow lights. The systems use far less water compared with conventional farms, require no pesticides or conventional ag chemicals, and can grow about 30 percent faster than crops on outdoor farms.

In 2018, Amazon's Jeff Bezos was one of several investors who pumped $200 million into Plenty, a San Francisco–based vertical farm start-up that aimed to bring its technology to China. Matt Barnard, the company's young CEO, has said that by 2020 he'll have built three hundred indoor farms in or near major Chinese cities.

At first, indoor food production struck me as a dangerously carbon-intensive, ludicrously costly, even postapocalyptic approach to growing vegetables. Agriculture relying on artificial lights requires, to begin with, a lot more energy to grow food than crops raised outdoors. The United States alone has more than 900 million acres of farmland, the vast majority of which will never end up under a roof. But in places with exploding urban populations, increasingly scarce water supplies, and limited arable land, this kind of ersatz farming is beginning to make more economic and environmental sense. Huge flows of investment are moving in this direction, and not just in China and Japan. One of the largest vertical-farming companies in the world is unassumingly headquartered in an old laser-tag building just a few miles outside of Manhattan in Newark, New Jersey. Inside, it feels like an industrial cathedral teeming with plants and processing power, a futuristic habitat designed to change the way we grow produce.

It sounds like science fiction, and to a degree looks like it, too, but these vertical farms aim to make local fruits and vegetables more widely accessible—cutting down on food waste and restoring freshness to produce that's otherwise trucked over thousands of miles. They manage at the same time to remove us from, and bring us closer to, the origins of our food.

CHAPTER 6

Altitude Adjustment

We must learn to think not only logically, but biologically.
—EDWARD ABBEY

FROM THE OUTSIDE, Jo-Ann Fabrics looks like any other strip mall store in downtown Ithaca, New York. Inside, it's a colorful, well-lit place that has the feel of a drafty attic. Its aisles are crammed with bolts of fabrics from calicos to cashmere. Within these aisles, Ed Harwood spent a good part of 2003, though he didn't blend in with the DIY moms who were the store regulars. Harwood has an absentminded-professor look—bespectacled, with thinning gray hair and a slight paunch. At the time he was frequenting Jo-Ann's, Harwood had a master's degree, a PhD, and a makeshift inventor's laboratory in an old factory near his house. "About my fourth time in the store, the clerks started eyeing me like I was half crazed. I'd be wandering the aisles and they'd ask me what I'm looking for. I'd always say I'm not sure—I'll know it when I see it."

At first Harwood was interested in felts. He sampled and tested at least a dozen of them, but found they were hard to wash and easily lost their shape. He moved on to the upholstery fabric aisles, buying swatches of sturdy polyesters and heavy linens, but the weave was too tight. Finally, several months into his regular visits to Jo-Ann's, he saw a bolt of oatmeal-colored cloth similar to the fleece used for jackets and baby blankets, "and it was like *ding, ding, ding!*" he remembers. This is the kind of fabric, he thought, that could trick a seed into thinking it was soil.

A few years earlier, in the late 1990s, Harwood had read in an agriculture trade magazine about an experimental concept of indoor farming called aeroponics—a system of growing plants in trays with their roots dangling in midair as they're fed a nutrient-rich mist. The systems used about 95 percent less water than conventional agriculture. "I'm going, wow! I had no idea plants could be grown with so little water," he recalls. At the time, upstate New York was in a serious drought, so water efficiency was on Harwood's mind, but unlike Jorge Heraud and Tony Zhang, he had no grand goal of saving the earth or its inhabitants. "I'm not a crusader type, the project just made practical sense to me somehow, even if I wasn't sure in the big sense *why*."

Harwood reasoned that aeroponics was a more efficient version of the hydroponics systems that have been used in greenhouses since the days of ancient Rome. But the concept had been applied only in laboratory settings to grow a few plants at a time. He wanted to grow thousands of plants, though there was no available information on how to do it—what material to grow the seeds in, or how to deliver the mist and simulate sunshine. There was also the problem that Harwood didn't know much about plants. He was a professor and administrator at Cornell University's College of Agriculture and Life Sciences, and his expertise was animals. He'd grown up south of Boston, where his father installed elevators in buildings and his mother ran a babysitting service. Since early childhood he'd spent summers visiting his uncle and aunt on their dairy farm in Vermont. He loved farmwork so much that by middle school he was boarding a bus to Vermont every Friday afternoon and returning home on the last bus Sunday night. Harwood was a mediocre student but loved tending the cows, and took notes on their behavior. By ninth grade, he'd moved in with his uncle and aunt and enrolled in high school in Vermont.

Ten years later, having gotten advanced degrees in microbiology and animal science from Colorado State University, he patented his first invention. Before graduating, he'd begun working at a start-up developing a system that identifies when cows are fertile. Harwood knew that cows become restless when ovulating, so he helped design a digital ankle bracelet that monitored the number of steps a cow takes. This was still the 1970s—so long before computers had made their way into mainstream agriculture "that the interface to our computer system was a ten-key pad." The digital bracelet is now used in dairies worldwide.

Harwood went on to get a PhD in dairy science with a minor in AI before he began teaching at Cornell; for fun, he also ran a popular satellite radio show on agricultural innovation. It was his research for the show that got him reading about aeroponics. Harwood began hatching a plan for his botanical experiments. He chose to grow baby greens mostly for economic reasons. "It hit me when I was at Wegmans supermarket," he says. "I looked at the greens for sale and saw that a pound of mature lettuce sold for a dollar, while a pound of the boxed baby greens sold for eight dollars—and took half the time to grow. I'm thinking, this can work." But when he shared the idea with colleagues, "everybody said it was nuts, way too expensive. That got me going. It really gets me going when someone tells me something won't work."

INDOOR FOOD PRODUCTION dates back to the Roman emperor Tiberius, who spent the last phase of his life in self-imposed exile on the island of Capri. Tiberius had a strange obsession with a pale-skinned, mild-flavored cucumber known as the snake melon, demanding that his kitchen gardeners serve him this summer delicacy every day of the year. His mandate spurred the creation of "specularia," wheeled carts that contained beds of soil that, as far as we know, produced the world's first out-of-season produce. This was before the existence of architectural glass, so the tops and sides of the specularia were made of a glazed, transparent stone.

Another millennium passed before the first glassed-in greenhouse structures were built, in the thirteenth century at the Vatican in Rome to grow tropical flowers and herbs that explorers of the time brought back from overseas, and it wasn't until the 1800s that greenhouse design got serious. The Dutch built greenhouses sophisticated enough to control not just heating but also light levels and humidity in order to grow medicinal tropical plants.

Modern greenhouses have grown a lot in size, volume, and sophistication, but also in environmental impact. Almería, Spain, has the largest concentration of greenhouses in the world—sixty-four thousand acres of contiguous growing chambers so collectively massive that their plastic roofs can be seen from space. The structures have been criticized for producing thousands of tons of plastic and agricultural waste, for damaging the region's water table, for exploiting low-wage immigrant labor, and for defying the local benefits of indoor agriculture—about 70 percent of the produce, which amounts to about $2.5 billion in production annually, gets exported to other parts of Europe.

There's one country that doesn't need these exports—Holland, which has produced a more sustainable model for indoor agriculture. Low-lying and flood-prone with soil poorly suited to farming, Holland sources fruits and vegetables from its own greenhouses designed with strict environmental standards to be water-efficient, minimally waste-

A patchwork of greenhouses in Almería, Spain

ful, and low-carbon. Some are capable of floating during floods; others have attached geothermal systems to regulate temperature. Japan, too, has been developing greenhouse technology post-Fukushima, in part due to concerns about radioactive soil. Outside of the United States, Japan and the tiny island nation of Singapore have built large-scale vertical farms—indoor growing chambers that, unlike greenhouses, allow no natural light. Today, indoor agriculture encompasses everything from the low-tech, passive-solar greenhouses not unlike the ones designed for Emperor Tiberius to highly sophisticated vertical farms, or "full-control agriculture," as Harwood calls it. Indoor production of vegetables in America has jumped more than 60 percent in just the last decade, due in part to the growing customer demand for local food, and to cost reductions in lighting and sensor technology.

David Lobell, director of Stanford University's Center on Food Security and the Environment, tells me that, in global terms, the rise of indoor agriculture is in part a response to the decline in arable land. The challenge of limited arable land extends well beyond China: drought, pollution, and erosion have wiped out "fully a third of the world's high-quality food-producing land over the last four decades," Lobell notes. Erosion, the biggest culprit, is the result of excessive ploughing and overuse of chemical fertilizers and pesticides that degrade the topsoil far faster than it can naturally regenerate. Given these growing constraints, the nations poorest in high-quality farmland are snatching it up in places with more favorable conditions and climates. China has purchased land in thirty-three countries, including Brazil, Ethiopia, and Argentina, in an effort to guarantee food security. The United Kingdom has bought farmland in thirty countries to shore up its supply. Germany, India, Saudi Arabia, and Singapore are also among the countries that have major investments in overseas farmland. Growers in the United States, despite the prodigious amount of American farmland, have holdings in more than twenty-five countries.

The trend, referred to by critics as the "global land grab," is expensive and complex, logistically and diplomatically. Indoor-farming advocates argue that it may become cheaper and safer, going forward, to grow fresh food locally in controlled environments rather than trying to chase and co-opt other arable land around the world. Especially as population and climate pressures intensify, the land-grabbing practice will likely get more challenging and dangerous.

HARWOOD DIDN'T IMAGINE his business would hit it big, or even hit it small, but he thought he'd do better than he initially did. Soon after he'd discovered the right growing cloth at Jo-Ann Fabrics, he rented out a work space in a canoe factory near Ithaca—a vast, dirt-floor basement crawling with salamanders, beetles, and millipedes he began to regard as his "officemates"—and built several hundred-foot-long rectangular steel boxes across which he stretched the cloth. He fed a hose with an attachment that mixed the water with a nutrient solution, a kind of homemade Miracle-Gro, and built a system of pumps and nozzles beneath the cloth that misted the chamber and moistened the ersatz soil. He scattered seeds from his favorite rocket plants onto the top of the fabric, flipped on some grow lights, and waited.

In two weeks, he had his first harvest of baby rocket, enough to fill about thirteen Ziploc bags with salad greens, which he brought around to his neighbors. He kept experimenting in his makeshift lab with different grow lights, adjusting temperatures to heat the cloth, trying different types of spray nozzles and nutrient solutions. Within a few months, he was able to grow hundreds of quarter-pound bags of baby greens per week. He'd branded his start-up "Great Veggies" and designed a logo on his desktop computer that he stuck on the plastic bags he used to package his lettuce. Two years into his start-up, he was regularly delivering lettuce to local restaurants and grocers, but he couldn't raise capital to scale up. By 2007, he shelved the business.

The following year, Harwood was working to get licensed to teach high school science when a guy named David Anthony, a private equity investor in Alabama, called him up out of the blue. "He'd found the old

Great Veggies website from a Google search—I didn't even know the site was still active—and said, 'I think you're the future of agriculture. I want half a stake in your company.' It was the biggest day of my career."

Anthony invested $500,000 and Harwood was back in business. He hired a couple of engineers, rebranded the company as AeroFarms Systems, and began building and selling indoor-growing equipment to farmers in the United States and the Middle East. In 2011, he got a call from two young graduates of Columbia Business School, David Rosenberg and Marc Oshima, who wanted to buy his company. Rosenberg had spent the previous decade building his start-up Hycrete, a producer of an ecofriendly weatherproof concrete, which he'd recently sold. Oshima had worked as a marketing executive at companies including L'Oreal, Toys"R"Us, and the gourmet food chain Citarella. Like Rosenberg, he knew a lot about business but nothing about agriculture. They bought the company, appointed Harwood chief technology officer, and started searching for industrial buildings in New Jersey where they could put the vertical-farming equipment to work.

By 2019, AeroFarms had raised more than $230 million in funding from investors including IKEA Group, Momofuku chef David Chang, and Meraas Group, a major firm in Dubai, and was growing greens in 70,000 square feet of urban real estate throughout the greater New York metro area. Harwood's equipment now grows dozens of baby green varieties—rocket, kale, mizuna, bok choy, and watercress among

David Rosenberg, Marc Oshima, and Ed Harwood

them—on aluminum towers rising thirty-six feet inside cavernous climate-controlled warehouses. AeroFarms churns out about seventy-five tons of leafy greens per month for urban supermarkets, restaurants, and cafeterias within a fifty-mile radius. "The growth is beyond anything I imagined," says Harwood, "but I do not presume to think that we're changing the world." For now anyway, AeroFarms is nothing more than a producer of fancy lettuces, he says, and therefore "still a heck of a long way from solving global hunger."

THE LARGEST VERTICAL farm in the United States is located at 212 Rome Street in the Ironbound section of Newark. "The future of farming is pink," Oshima tells me one sunny afternoon in March, as he and Harwood usher me inside the windowless building, which, in its former life, was a steel mill. The company is headquartered a few blocks away in a warehouse that was once Inferno Limits, a laser tag and paintball arena, the walls and ceiling of which are still covered with graffiti-style murals and splattered paint. The former steel mill is now also spangled with color: fuchsia light emanates from the towers that rise from floor to ceiling throughout its cavernous growth chamber. Filled with the bright smell of vegetation and humming with the sound of pumps, misters, and fans, the place looks more like an Amazon packaging center than a farm.

The greens are barely visible within the towers, each holding foot-deep growing beds stacked twelve stories high in rows eighty feet long. People in hazmat-looking jumpsuits and hairnets are shuffling around the concrete floor quietly, peering into the glow of their tablets and the trays. To reach the higher trays, they rise up on mobile cherry pickers.

"Historically farming has been about adapting plants to the environment. Vertical agriculture is about adapting the environment to the plant," says Harwood. "It may look unnatural to you, but from the perspective of a plant, all this is entirely natural—they're getting only and exactly what they need."

Oshima describes the operation, just as Jorge Heraud and Tony Zhang describe theirs, as technology with an ecological conscience—innovation that supports rather than competes with goals of sustainability. Others in the vertical-farming industry have described the practice of indoor agriculture as "post-organic"—a type of food production that's pesticide-free and uses a small fraction of the water and fertilizer required on outdoor farms. It's also entirely insulated from climate volatility.

The AeroFarms lighting technology is similar to the systems used to grow plants on the international space station. The sun is replaced by LEDs that emit only a narrow band of high-intensity blue- and red-spectrum light, giving off the Pepto-Bismol hue. A cloth similar to the one Harwood found at Jo-Ann's is stretched across the top of the trays. The plant roots dangle below the fabric in midair like feathery icicles, drinking in a high-pressure, nutrient-rich mist within the trays. Cameras and sensors track and monitor the progress and needs of the plants in meticulous detail, collecting and analyzing thousands of data points that guide the care of the greens as they grow.

The growing process begins with the automatic seeder. The seeds are scattered by a guided mechanical arm, using imaging software and algorithmic analysis to produce the perfect growing arrangement across the cloth, which can be removed, scraped, washed, and reused after each grow cycle. The seeds rapidly germinate—in less than half the time it takes for seed germination in the field. The trays are stacked into the tower with the warmth-loving lettuces at top and cooler varieties at bottom to account for the rising heat.

The seedlings bask in their trays like so many thousands of lazy sunbathers in a giant tanning bed under the pink grow lights. There are advantages to LED lights, Harwood tells me: because they produce no radiant heat, they can be placed directly overhead; the plants don't expend energy growing upward and developing stalks—instead they grow outward into leafy material. Aeroponics is more expensive, complex, and glitchy than hydroponics, but has a big advantage: because the roots aren't submerged in water or soil, they're exposed to higher levels of oxygen so that the plants grow faster.

AeroFarms further speeds the growing process with extra carbon dioxide, which the plants breathe in during photosynthesis. The ambient air in the warehouse is filtered, ventilated, heated, and cooled. Tanks of CO_2 enrich the concentration inside the building to a thousand parts per million—more than double the average ambient CO_2. "Here's the brain of the grow tower," Oshima says, opening a large metal box filled with a snarl of wires. "We have cameras and sensors all over the place—tens of thousands of sensing devices interpreting millions of data points at any given time." The data points are related to the plants' growth process and all the variables that affect it—among them, temperature, humidity, spectrum and intensity of light, nutrients absorption, and oxygen and CO_2 levels.

The "brain" is governed by an AI system—sort of a student version of Heraud's—programmed with thousands of images that show what a normal plant should look like at each stage of growth; based on these standards, the cameras can automatically detect aberrations in the color, shape, and texture of the leaves. If a camera notices irregular growth, the system alerts scientists via apps on their phones, and they can then troubleshoot the problem remotely—adjusting the conditions accordingly. "The goal is always more uniformity," says Oshima. The crop data is stored in a cloud-based system that's continually mined by the company's operations team, food safety team, finance, and R&D, "bringing more knowledge and control to all aspects of the business."

Environmentalist Paul Hawken supports vertical farms but doesn't see them moving beyond niche applications: "There is enormous value to urban food production and vertical farming, but the latter will not reduce greenhouse gas emissions, nor will it alone feed the population. There are costs to mechanized agriculture, whether indoor, outdoor, or vertical. When you maximize speed, productivity, and uniformity, you lose the precious connection between plants and the places where they're grown, and the people who grow them."

Highly controlled agriculture is also high-risk. Because there's no soil or other barrier to protect the roots, even a small amount of bacteria, mold, or other contamination in the root chamber can harm the plants. And because the plants are getting the exact amount of water and nutrients they need to survive, any slip in the system—say, a pump or sprinkler or timer breaks down—will cause the greens to suffer. If there's a power outage and the tray reservoirs stop filling with mist, the greens could die within an hour. "Aeroponic plants are like the boy in the bubble," Caleb Harper, director of the Open Agriculture Initiative at the MIT Media Lab, tells me. "They're fine so long as the bubble doesn't pop."

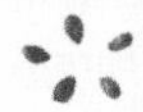

A MONTH AFTER AeroFarms opened its farm at 212 Rome Street, one of the world's few established vertical-farm companies at the time, Atlanta-based lettuce producer PodPonics, tanked. It wasn't long after the company had gotten an offer from Kroger to order $25 million of its lettuce annually if it could build the farm to support it. "This was our wildest dream, we were ready to go, this was everything we wanted," said former PodPonics CEO Matt Liotta at an industry conference soon after he shut down the business. "And then we realized how much capital this was going to require, how many people we were going to have to hire. We were simply incapable of building everything they wanted." Liotta added at the end of his commentary, "This is really a manufacturing game. It is not an art. If you want to do art, get a garden."

Soon after PodPonics was shuttered in 2014, another smaller but high-profile hydroponic lettuce-growing outfit, FarmedHere, closed its 90,000-square-foot facility in Chicago and canceled its plans to invest millions in a new farm in Louisville, Kentucky. High labor and energy costs meant it needed to sell huge volumes of produce to break even, and scaling fast was too risky. Around that time, Google Ventures, Toshiba, and Mitsubishi, all of which had been investing in urban-farming ventures, pulled out of their investments on the grounds that the technology was too young. "It's what happens when industries are born—only one of a hundred early efforts will survive," says Harwood.

Yet more and more players are taking on those odds. Some, like Brooklyn-based Square Roots, founded by Kimball Musk, brother of Elon, are producing their greens in shipping containers with LED lights hung vertically like curtains between screens of crops. In 2018, a few years after getting out of the vertical-farming game, Google Ventures led a $90 million investment round in Bowery Farming, another heavily digitized operation producing hydroponic lettuces based in Kearny, New Jersey. The $200 million bet on Plenty from Jeff Bezos and other investors was the biggest injection of funding the industry had yet seen, but Plenty is using hydroponic systems, a technology that requires more water than aeroponics—a potential disadvantage long term.

AeroFarms has turned down multiple offers of eight-figure investments, says David Rosenberg, because he believes in more gradual growth. Even after three years of product sales, the company is still barely turning a profit. "We're playing the long game—our profits will grow with scale," Rosenberg tells me. In the meantime, though, the only other notable large-scale aeroponics effort is at NASA, in its "off-world agriculture" laboratories.

Inside 212 Rome Street

Economic challenge number one for any vertical-farm company is still energy demand. A study by Harwood's old colleagues at Cornell University found that "hydroponics offered eleven times higher yields but required eighty-two times more energy compared to conventionally produced lettuce." Harwood contests this data, saying that it doesn't account for the costs of cold storage, transportation, and pesticide impacts. He adds that aeroponics' increasing efficiencies with lighting have gone a long way to reducing costs.

AeroFarms has designed a lighting system, Oshima says, that's "far more nuanced than blending red and blue," but he wouldn't get more specific. Future cost savings are inevitable, he explains: between 2012 and 2014 alone, LED lighting efficiency jumped about 50 percent, and by 2020, it's expected to climb another 50 percent as costs drop. AeroFarms is also experimenting with strategies of modulating the light intensity throughout the growth cycle, giving plants only the amount they need. "You don't just flip a switch; the light needs of each variety of greens vary day to day and sometimes hour to hour from germination to maturity, as does the length of time each needs to rest from light exposure," Oshima says. "We can manage lighting with exquisite detail using sensors, cameras, and machine learning."

When I press him on the carbon footprint of sunless agriculture, Rosenberg says that AeroFarms has chosen to locate some facilities, like the one in Buffalo, New York, for example, next to a hydroelectric plant that provides zero-carbon energy. His Newark, New Jersey, facility is powered autonomously with geothermal and an on-site natural gas turbine; he's experimenting with CO_2 capture to pipe the gas into the growth chambers and accelerate plant growth. As efficiencies in solar improve, he'll also pursue on-site solar "so our plants are grown indirectly, if not directly, by the sun."

Rosenberg also makes a compelling case for the natural carbon sinks that can be grown on the agricultural land displaced by urban farms. "Our facilities are 390 times more productive than traditional field farms on an annual per-acre basis," he observes. "Think about returning the acreage we're offsetting to its natural, pre-agricultural form—growing trees there, restoring the land to wilderness that breathes in CO_2. What could be better for the planet than that?"

ANY SUN-HUNGRY CROP grown on a mass scale that stores well, like wheat, corn, and rice, is—and, in all likelihood, will always be—a poor candidate for indoor farming. Hypercontrolled growing conditions make sense only for crops like leafy greens that are at once highly perishable, nutrient-rich, and climate-sensitive. Leafy greens can be grown and harvested quicker than any other produce, generating immediate cash flow in a business that requires a lot of investment up front. Harwood's machines can produce about twenty-five to thirty harvests annually—transforming a dormant seed to a harvestable plant in twelve to sixteen days, depending on the green variety. That same crop in an outdoor field, using the same seed, would take about thirty to forty-five days to reach maturity, producing less than a quarter of the yield.

It's an advantage, too, that you can sell virtually every part of the leafy green that you grow—meaning no energy is wasted growing extraneous parts of the plant. "Light costs money, so growing, say, an avocado tree—trunk, bark, branches, leaves, fruit—under LED lights to only sell the fruit doesn't make sense," says Harwood.

Lettuce is also a great candidate for indoor farming because there are so many problems with growing it outside. The cool, dry, ninety-mile-long Salinas Valley south of San Francisco produces more than two-thirds of the nation's leafy greens. Most of the remaining third comes from Yuma, Arizona, where lettuce grows during winter. "This stuff is perishable, precious, and tremendously expensive to move it across the country, and the products lose value along the way," says Oshima. "That's our arbitrage opportunity—coming in with a better, fresher product without the expense of chilled transportation and a long supply chain."

One industry analyst I spoke with called leafy greens the "bottled water of produce" because they require a disproportionate amount of energy, water, and labor to grow—only to have most of them thrown away from rot. One head of romaine lettuce takes about 3.5 gallons of water to cultivate—a significant statistic, especially in regions facing escalating water shortages. Lettuce also gets a bad rap for carrying food-borne illnesses. In just the last decade, bagged spinach producers, including the organics giant Earthbound Farm, have sold products contaminated with *E. coli* from cow manure that have killed five Americans and diseased hundreds. In 2018, fifty-two people across fifteen states were sickened by *E. coli*–tainted romaine.

The only foolproof option is a chemical wash—or growing the greens indoors. "You look at all the vulnerabilities—contamination, climate sensitivity, nutrient erosion, and waste. Lettuce is a category that needs transformation," says Oshima. "We have the technology advantage."

The baby romaine lettuces I see harvested at 212 Rome Street are so thick and glossy and rippled they look like small glowing green brains. And while I'm expecting lettuce that has never been exposed to sun or soil to be limp and tasteless (as is often the case with hydroponic lettuces), instead the flavor is bright and fresh and the texture is firm.

In the early days of his enterprise, Harwood assumed that the biggest advantage of indoor aeroponics systems would be the water efficiency, the potential crop volumes, and the proximity of local markets. "What we've discovered is that even bigger value is the data we're collecting on how plants behave and exactly what they need to thrive." That knowledge gives AeroFarms the ability to coax the plants to grow with specific attributes.

"*Terroir*" is a French term often used by sommeliers to refer to the special environmental and climatic conditions—humidity levels, heat stress, air quality, oxygen levels, soil quality, even the mineral composition of the water used to irrigate the vineyard—that together produce the grapes for a particular vintage of wine. It's the set of all environmental factors in, say, Bordeaux, France, that affect a crop's phenotype—colors, flavors, acidity, textures, odors, and so on. But it's not just wine; many other crops are studied for their *terroir,* including coffee, tobacco, chocolate, chili peppers, hops, agave, tomatoes, and cannabis—high-value crops with complex flavors and sensory impacts. And with its digitized growing environment for salad greens, AeroFarms hopes to control a kind of digital *terroir* that in turn affects the flavors and textures of its lettuces.

Paul Hawken finds this dubious. He thinks it's a long shot, at best, that humans will ever fully understand *terroir.* "Remember that the most complex ecosystem in the world is a square inch of soil. The interactions happening within the microbiome and between the soil and the roots of plants, between the plants and the weather and the geological history of the land—that's what creates amazing flavor; it's the soul of food—it's beyond algorithms, beyond comprehension."

Harwood counters that his team's ability to understand the factors that influence a plant's flavor is getting ever deeper and broader. "We can isolate for very specific variables. We can say, Hey, with lower humidity or with higher heat at this time of development, and all the other variables the same, that affects a certain characteristic"—it could be the color of a lettuce leaf, or the sweetness in a strawberry, or the amount of lycopene in a tomato. And over time, as they develop a vast body of data on the characteristics they're looking for, they're better able to control them.

The future for AeroFarms, says Oshima, is homing in on value-added attributes for crops. "We can work with a particular chef, for example, to develop produce to his or her specifications—a lettuce that's spicier, or more red, or more serrated, or sweeter." Adjusting plant characteristics in this scenario could become as simple as tweaking a photo with an Instagram filter—but it's still in the early stages. "We're just beginning to explore this," he says, "but as we build our reservoirs of data, yes, we can get that specific, down to levels of macronutrients and micronutrients, because we're able to monitor real-time the absorption rate of the nutrients as they're taken up into the roots of the crops. In the field, you don't have nearly that level of control."

AeroFarms is developing machines that can grow a range of high-value produce, including berries, tomatoes, grapes, cucumbers, beets, and other root crops. Some of these crops will be grown, like the lettuces, for retail, while others will be grown for their by-products. In an R&D farm near the 212 Rome location, the AeroFarms team is collaborating with multinational food companies on plant-based research, exploring, for example, how plants can function like tiny organic machines that can manufacture natural flavors, dyes, and added nutrients for processed food products.

Greens growing in an AeroFarms tray

While this kind of controlled crop-testing is a new phenomenon in the food industry, it's not so new for cosmetics and pharmaceuticals, and Harper of MIT sees potential in applying computer-aided precision to specialized food production. At the Media Lab, he's developing "personal food computers"—essentially, smaller indoor growth chambers that could work at the household level. These PFCs would tap into a database for what he calls the "climate recipes" necessary to grow a particular kind of fruit, vegetable, or plant extract. "What we need is an open-phenome library so that anybody, anywhere, can pull down a climate recipe they need to grow whatever plants they want," Harper says, whether it's the basil used at Momofuku, the high-lycopene tomatoes of southern Italy, or the serrano peppers of Mexico. "The tricky thing is that right now there exists no baseline programming language for controlled-environment farming, no Linux for agriculture. How do we create a common syntax so that growers everywhere can share data about their growing conditions? That will be the basis for the internet of food."

The reason we've never done this, says Harper, is that the priority for American farmers has been almost exclusively yield. "We've spent a long time optimizing our agriculture for two things—cheaper food and more of it—at the expense of quality. I'm not knocking what came before us. I think the Green Revolution was great. It made a lot of lives possible. But now we can do more, far better."

THE PREMISE OF the 1972 film *Silent Running* is botanical apocalypse. The story takes place in a greenhouse floating through space. We follow Bruce Dern playing ecologist Freeman Lowell on his tragicomic quest to preserve plants for future humans. Assisted by three quirky robots, Lowell revolts when his human superiors decide to destroy the greenhouse to make room for cargo. Lowell and his bots go to white-knuckle extremes to keep the greenery alive.

I saw the film when I was twenty-three and found it funny and absurd in equal measure—a send-up of the eco-utopic notions of the early seventies. Lowell is like a post-Armageddon Saint Francis strolling next to trickling man-made streams, caressing plants in his biosphere while birds alight on his shoulder and Joan Baez sings in her trademark vibrato about "rejoicing in the sun" and nature being "your precious child." It's end-of-nature meets back-to-the-land.

Twenty years later, as I'm strolling through the AeroFarms warehouse—a botanical Ark in its own right—past acres of leafy greens in their pink tanning beds, I realize I'm not far removed from a scene in *Silent Running*. The 1972 sci-fi lampoon wasn't that ridiculous after all. I left the AeroFarms facility at 212 Rome with a desire to weed my shabby little garden, to dig around in some soil and eat the mottled, misshapen things that grow in it. I don't particularly like the idea of eating lettuce or strawberries that have been algorithmically optimized for uniformity. Nor do I like the idea of losing the intimate connection between plants, land, and people. But I do like the notion that putting more food production indoors might shrink long-distance food chains, reduce rot and waste, and liberate farmland from ploughing and chemical inputs. I like the idea—however fantastical—that the land could be restored to carbon-absorbing forests, reversing some of the environmental impacts that agriculture has had over the past ten millennia.

In the nearly half a century since *Silent Running* was made, a lot has changed. We haven't just lost a third of the world's arable land, we've accelerated the rate of loss going forward. Much of that loss is occurring in politically vulnerable regions. Saudi Arabia, for example, imports 75 percent of its food. Other desertified countries in the Middle East, including Iraq, Qatar, and the United Arab Emirates, also heavily depend on imports. As Caleb Harper puts it to me, having just returned from Abu Dhabi, "Their soil contains like 0.001 percent biologic material. They don't have any rivers. They're not getting any more water any time soon. So indoor farming that relies on less than 10 percent of the typical water supply—that can be their version of twenty-first-century victory gardens."

The world's population is expected to reach 9.8 billion by 2050—33 percent more people than are on the planet today—according to projections from the United Nations. About two-thirds of them are expected to live in cities, continuing an urban migration that has been under way for years. Today just over half of the global population lives in cities. By 2050, fourteen of the world's twenty biggest metropolises will be in Asia and Africa, with Jakarta, Manila, Karachi, Kinshasa, and Lagos joining Tokyo, Shanghai, and Mumbai. By then, given population growth and changing diets, the United Nations estimates we'll need to raise global food output by 70 percent from 2009 levels. It stands to reason that eventually we'll need to grow our food upward as well as outward on a bigger scale.

But Ed Harwood, Caleb Harper, and other advocates of vertical farming aren't proposing to replace conventional agriculture. They're proposing to decouple the production of commodity crops like wheat, corn, rice, and soy—the stuff that stores well—from the stuff that doesn't: fresh produce. Harper explains, "Ultimately, what it looks like in about fifty years is that at least 70 percent of crops will continue to be grown conventionally, on large-scale, centralized farms." A portion of the rest—the fresh vegetables and fruits that suffer most from loss of flavor and nutrients in transport—can be produced in local outdoor farms, community vegetable gardens, and vertical indoor facilities for growing urban markets.

In order to make this vision work, says Harwood, we'll need a lot more farmers—and a different breed of farmer than Andy Ferguson in Eau Claire, Wisconsin. Harwood stresses that growing sustainable food will increasingly require dynamic teams of people with the kind of technical skill sets that young urbanites may be more likely to have than seasoned heartland farmers. Like Jorge Heraud's Blue River, AeroFarms is more populated with software hacks than green thumbs. "This kind of growing requires horticulture, biochemistry, mechanical engineering, electrical engineering, programming, food safety specialists, architects, and so on," Harwood tells me.

Going forward, such a diversity of talent may be as necessary for the production of produce as it is for proteins. Consider fish farming, aka aquaculture. As environmental pressures and overfishing threaten wild fish populations, aquaculture is becoming a linchpin of protein production worldwide, albeit a controversial one. On the next leg of my journey, I travel to Norway to meet a salmon farmer trying to forge a third way approach to large-scale aquaculture using old-, new-, and otherworld tools and techniques.

CHAPTER 7

Tipping the Scales

The charm of fishing is that it is the pursuit of what is elusive but attainable, a perpetual series of occasions for hope.

—JOHN BUCHAN

ALF-HELGE AARSKOG is pacing. Back and forth he walks across the polished hardwood floor of a barge anchored in the crook of a fjord on the western coast of Norway. It's November and the sky is cloudless, the mountains sleek and snowcapped, the water around us a clear sapphire blue. The barge sits alongside one of the world's largest salmon farms, with eleven circular underwater cages nearby that collectively hold about four thousand tons of fish, or $60 million worth of seafood. The main room of the barge has the feel of a trendy hotel lobby with its blond wood panels and spare Scandinavian design. It's furnished with leather couches, a conference

A Marine Harvest salmon farm

table, and exercise equipment. On one wall are huge monitors streaming video from the cages. Aarskog wears black from head to toe and has thick, scowling eyebrows, spiky brown hair, and a long, sharp jaw. He scans the footage—masses of salmon swimming in circles like glittering cyclones—and mutters what I take to be Norwegian profanities.

Aarskog is the CEO of Marine Harvest, the world's largest salmon fishery, with 220 farms in Norway, Chile, Scotland, the Faroe Islands, and Canada. Aarskog supplies his fish, a third of which are certified "sustainably farmed" by the Aquaculture Stewardship Council, to companies like Whole Foods and Walmart and to restaurants worldwide. Aarskog has come to this farm, near the island of Frøya, to meet his operations manager, who recently reported some bad news. Underwater sensors showed that the feed consumption in some of the cages has been declining when it should have been rising. A sample of fish pulled at random from the cages has confirmed the worst-case scenario: every fish on the farm is in immediate jeopardy.

Many threats exist in ocean aquaculture. A plague of jellyfish can wipe out a cage of fish in one fell swoop, an algal bloom can starve the fish of oxygen, a breach in the cage netting can result in a mass escape. This problem's worse—and almost imperceptible. Lurking among the millions of captive fish is their tiny nemesis: *Lepeophtheirus salmonis,* aka the sea louse. Mature sea lice are gray, lentil-sized crustaceans, also called ectoparasites, that look like tiny fanged tadpoles. About a dozen sea lice can kill a fish, clinging to its scales as they consume blood and flesh. The lice have coexisted with salmon in the wild for millennia, but only recently have they become a serious threat.

Aarskog, who is fifty, has been fighting sea lice for more than thirty years. His love of fishing came from his grandfather, a hunter and angler who took him fishing on weekends as a boy and taught him how to sell catch at the local market. Aarskog got his first job at a salmon farm when he was fourteen and the industry was still in its infancy. He built cages from wood planks that could hold a few thousand salmon each; he dumped vats of sardines and anchovies into the cages to feed the salmon; he cleaned the "mort tanks" holding quarantined fish that were diseased and dying. His boss also had him slice thousands of onions and garlic cloves into cages as a hopeful sea lice deterrent.

The lice problem was limited then because the industry was small. The cages held a fraction of the 200,000 fish typically farmed per cage today. The larger and more concentrated a population of captive fish, the bigger the potential plague. Like the superweeds that resist herbicides in huge crop fields, growing populations of "superlice" are now defying nearly every effort to subdue them. The industry, in a sense, has created its own monster.

The lice outbreak in Frøya is still in early stages. We see no more than a few parasites on the several dozen fish that are pulled from the cages, and many are clean. The salmon looked perfectly healthy to me, yet there's no choice but to harvest the salmon now—before the sea lice population explodes. Each female louse can produce two strings of a thousand eggs, which means infestations spread quickly from cage to cage and can threaten wild fish migrating through nearby waters. The salmon in Foster's cages currently weigh three kilos each, only about 60 percent of their full-grown weight. That's a loss of about $24 million, he tells me.

Marine Harvest can easily absorb this cost on a single farm, but the lice have spread to most of Aarskog's farms. Between 2015 and 2017 his total harvest across 220 farms declined by 12 percent. Some of his competitors have been even harder hit. The problem, he says, has reached the "status of a nightmare" in an industry otherwise poised for massive growth. Aarskog and other industry executives have banded together to mount a technological arms race against their microscopic foe, investing hundreds of millions into possible solutions, some of which seem as strange and improbable as the parasites themselves.

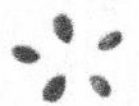

SALMON ARE ANADROMOUS, meaning they're born in freshwater rivers and travel long distances to saltwater seas to feed before migrating back upriver to spawn. For thousands of years, the wild salmon species of Norway have used the fjords for safe passage between their rivers of origin and the Norwegian Sea. Today the fjords also contain much larger populations of farmed fish—about 1.5 billion Atlantic salmon are grown here a year. Norway is to farmed salmon what the United States is to beef—by far the biggest producer. Salmon farming is a $14 billion industry worldwide, according to Aarskog, and has doubled in the last decade. Norwegian production is high in large part because the cool, calm fjords are protected from the winds and currents that make ocean aquaculture difficult. This region hasn't yet been affected by warming trends that have begun to harm other salmon habitats around the world.

Aarskog, whom I've heard described with both derision and respect as a modern-day Njord, the Norse god who tamed the seas and controlled the region's prosperity, commands about a quarter of the global salmon farming industry. When I mention the sea god comparison to Aarskog, he bristles. "Don't call me a visionary," he implores. "I am not the king of anything—no one can be."

Alf-Helge grew up on a sheep farm in Alesund, a fishing town on the west coast of Norway south of Frøya, and started working at a young age. By six, he was herding sheep up mountains to feed them grass on the high pastures. By eight, his father had taught him how to slaughter sheep by himself. "I was working long hours before it is probably legal to work. What do you call it in America? Child protection services? They would have been visiting my parents," he deadpans. I take his comment as a joke, though it's hard to tell with Aarskog. In the many hours I spend with him during my visit and by phone thereafter, he laughs exactly once, when I comment that for a guy who runs a $3.2 billion company, he's young (he was forty-nine when we first met in late 2016). "Young!" Aarskog howls. "I feel very old."

The man is obsessive about physical fitness. He trains daily and competes every four months to test himself in three disciplines: terrain running, mountain biking, and cross-country skiing. "My goal," he tells me, "is that when I'm eighty I'll be able to beat my time when I'm sixty-five—and not the way Americans do it, with performance-enhancing drugs." Alf-Helge governs himself with a rigorous work ethic and eschews the trappings of luxury. When I ask him if he likes to fly-fish the rivers of Norway, he rejoins, "Fly-fishing is for rich people. I come from a farm." While Aarskog's wife drives a Tesla, he's an Audi loyalist who scoffs at the electric car trend. He considers it a distraction from the real climate-change culprit: meat. "It's completely crazy that there are so many tax incentives for people to buy electric cars, but none to convert people away from meat-based diets," he tells me.

Aarskog's blend of humility and pragmatism hits me like the Nordic air—bracing and not exactly pleasant, but good for the constitution. There's no question that he lives a life closer to nature than I ever have or will; in addition to all that terrain running and mountain biking, he spends weekends and summers in a cabin on a tiny coastal island completely removed from commerce of any kind. He's also far less sentimental about the environment than I and other backyard-gardening, climate-fretting people are, and it occurs to me throughout our conversations that maybe that's a good thing: maybe he possesses the right combination of both detachment from and connection to nature that's needed to solve major systemic problems with aquaculture going forward.

For all his humility, Aarskog is bullish. Setting the pace for the

Alf-Helge Aarskog

industry, he's more than doubled Marine Harvest's fish production in the last decade. The growth has brought gains for a fish-loving, middle-income consumer like me: the prices of his salmon products—smoked, frozen, and fresh—have come down 20 percent since 2005. Aarskog's fish are sold in supermarkets worldwide, including at my local Kroger. He also produces those premade sushi boxes you see in airports and malls. On the subject of his competitors, Aarskog doesn't mention anyone in the fish space—he already dominates it. "It's Tyson Foods, it's Smithfield, the big meat producers—that's the market share I'm after," he tells me, reiterating the potential climate-change benefit.

Aarskog plans to grow his company tenfold by midcentury to help mobilize a "Blue Revolution," as he describes it, "in which aquaculture will replace wild-caught fish and provide sustainable protein to billions of people." Oceans are 70 percent of the earth's surface but provide only 2 percent of our food, Aarskog tells me more than once. "That has to change."

Some critics balk at this. Seafood is already the dominant protein source for three billion people, mostly in Asian countries, but whether a farmed supply of it can be sustainable on a global scale is the subject of much debate. "Industrial aquaculture, especially for salmon, is inherently unsustainable," says Don Staniford, director of the Global Alliance Against Industrial Aquaculture (GAAIA). "We should end salmon farming. Period. Full stop." The environmental costs of salmon farming go well beyond lice. The management of salmon waste, cage escapes, and the quantities of wild fish that are killed to feed salmon are still big challenges. Salmon farming today represents only about 5 percent of the global aquaculture industry—cheaper fish like tilapia, carp, and catfish are produced in much larger quantities, largely for Asian markets—but it's the fastest-growing sector and by far the most profitable. "This is where the largest investment in R&D and innovation is happening," says Josh Goldman, owner of Australis Aquaculture, a Massachusetts-based start-up. Goldman has benefited from advances in salmon aquaculture even though he raises barramundi, a tropical whitefish about half the size of a salmon. "The use of energy-dense fish feeds and underwater cameras that have been critical to improving efficiency—we owe that to the salmon industry. From a tech standpoint, where salmon farming goes, so go many other kinds of aquaculture."

Aarskog has signed on to goals set by environmental groups to create a zero-lice, zero-waste, zero-escapes company. He's also planning to eliminate wild fish from his fish feed altogether. "Marine Harvest has come a long way," says Ingrid Lomelde of the World Wildlife Fund's Norway office, "but growing this company at breakneck speed while meeting these goals will be, at the very least, difficult."

When I arrive in Norway, I'm curious to investigate the challenges of large-scale aquaculture, but truth be told, I'm also hoping for good news about the frontier for the simple reason that we need it. Wild populations of almost every fish species, including salmon, are declining across the globe because of overfishing, climate change, and other environmental pressures. According to United Nations data, the global demand for seafood will grow at least 35 percent in the next two decades, given current population and economic trends. My travels in Norway manage to dispel many of the concerns I'd brought with me about salmon farming, and to reveal others I hadn't begun to consider.

I would come to see that aquaculture, like genetic engineering and vertical farming, is another fast-growth industry with huge potential and almost equally big risks.

IN 2016, JUVENILE Chinook salmon were cooked by the millions in California—and not by ovens or stovetops, but in the Sacramento River. The waters of that 450-mile-long river, which are fed by snowmelt running off the Sierra Nevada mountains, had been getting shallower and hotter throughout the previous years of drought—too shallow and hot for juvenile salmon to survive. The number of young Chinook salmon that survived the journey through the river to the Pacific Ocean declined from more than 4 million in 2009 to fewer than 300,000 in 2015. Peter Moyle, a UC Davis fisheries biologist who authored a study titled "State of the Salmonids: Fish in Hot Water," predicts that fourteen salmon species native to California, including Chinook, will be extinct within fifty years, given current climatic trends.

Wild salmon overcome tall odds during their migration between rivers and oceans, which can be hundreds or even thousands of miles long. They battle up rapids and waterfalls, over power dams, and past fishermen's nets; they defy predators—eagles, otters, bears, and humans—along the way. "Salmon are an unusually hearty and robust fish that can get past just about any obstacle," says Moyle, "except increased temperatures—especially at the juvenile stage, they have limited ability to adapt."

Warming trends are affecting many other types of fish and shellfish around the world. Erica Goode reports in the *New York Times* that "two-thirds of marine species in the Northeast United States have shifted or extended their range as a result of ocean warming, migrating northward or outward into deeper and cooler water." Lobster populations, to take another example, are shifting from southern New England toward Maine. The cod fishery in New England has nearly collapsed because of the rapidly warming waters of the Gulf of Maine, which is rocky and shallow; the young cod have declining food sources, and as they move to deeper waters they face more vicious predators. Black sea bass, scup, yellowtail flounder, mackerel, herring, and monkfish, among many other species that for centuries have thrived off the east coast of the United States, are migrating north. "The center of the black sea bass population, for example, is now in New Jersey, hundreds of miles north of where it was in the 1990s," writes Goode.

Warming oceans are also acidifying—a phenomenon sometimes called the "evil twin of global warming." Acidification is caused by the continual decrease in the pH of the earth's oceans as a result of CO_2 absorption into the water from the atmosphere. Acidification is hard on shellfish, especially oysters and crabs. Scientists predict the Dungeness crab population, for example, will decline 30 percent in the next few decades due to acidification, disrupting a $180 million industry in America.

"It's crisis time for wild fisheries," says Australis's Goldman. One advantage of aquaculture is that it can control conditions that in the wild are going out of whack. Goldman continues, "By all accounts, feeding humanity in the era of climate change will require significant increases and major advances in farmed seafood."

Yet for now, it's aquaculture itself—and salmon farming in particular—that's posing some of the most serious threats to wild fisheries. "It's more the problem than it is the solution," says Don Staniford. There are, to begin with, literal ways in which salmon farms are consuming wild fisheries. Salmon are carnivores that can eat many times their weight in anchovies, herring, squid, eels, shrimp, krill, and other wild marine life. "You can't grow an industry that depends on wild feedstocks that are themselves declining," says Ingrid Lomelde. Farmed salmon don't just feed off wild fish populations, they pollute their habitats. Up to 70 percent of the food fed to a farmed salmon will end up back in the water in the form of urine, feces, and uneaten feed. "For decades the salmon industry has used the ocean as an open sewer," says Staniford. "The excrement settles on the ocean floor and chokes the marine ecosystem in that area."

Farmed salmon can also genetically contaminate the natural ecosystem. When the farmed fish escape and breed with wild salmon, chances are high that the offspring won't have the genetic hard-wiring necessary for survival. Farmed varieties of salmon are genetically distinct from their wild cousins. They lack the homing instinct to return to their river of origin because they were born in commercial hatcheries. And because they face no predators in a cage, they lose their genetic instinct to be wary of danger. "Farmed fish are good at fighting for food pellets—that defines success in a cage, but it doesn't do you much good in the wild," says Staniford.

Even without escapes, farmed fish can pose a threat. The feed in the cages attracts the wild salmon that are swimming past the farms. Wild juveniles can slip inside the nets and get cannibalized. Those that aren't eaten can pick up diseases and parasites infecting the farmed fish and take them into the wild. In the early 2000s, farmed Atlantic salmon spread a disease known as ISA (infectious salmon anemia virus), thus devastating wild Pacific salmon populations. In the 1990s, salmon farms spread SAV (salmonid alphavirus), causing pancreatic disease in wild populations throughout Europe. Eventually, Marine Harvest and other industry leaders developed vaccines to control these threats.

Sea lice have proven much harder to manage than the disease outbreaks. "For now, lice is the most pressing challenge—both to the industry and to the wild salmon picking up the parasite," says Lomelde. She estimates that sea lice kill about 50,000 wild salmon a year. Unlike Staniford, though, she sees the industry on a positive environmental trajectory—"two decades ago it was a young cowboy industry with little regard for the environment. That's changed. They realize they can't grow if they don't do it sustainably." In general, Aarskog maintains, whatever's bad for wild fish populations and the marine environments in which he builds his farms is also bad for farmed fish. "We don't want them living in lagoons of waste, they can't grow well, we don't want them escaping, and we certainly don't want them infected with parasites," he explains. "We are highly motivated to eradicate these problems."

Aarskog is approaching the mission of lice eradication with all the enthusiasm of *Caddyshack*'s Carl Spackler in his battle against the gophers. "We'll do whatever it takes," Aarskog tells me. "There are dozens of new concepts in play across the industry, and sooner or later a combination of them will work."

WE'RE SHIVERING ON the rim of one of Aarskog's salmon cages under a setting November sun. Thousands of fish are circling the top of the cage, jumping and thrashing as they jockey for feed pellets that are blowing like confetti from a suspended plastic hose. The fish don't seem to notice that it's 20 degrees below zero Celsius outside, or that among them, puttering darkly, is a mechanical novelty: a robot that looks like an oblong R2-D2 shooting green laser beams in all directions.

This device, dubbed Sting Ray, was built by deep-sea oil industry engineers specifically for sea lice extermination, and is one of the more eccentric weapons Aarskog is testing in his arsenal. Sting Ray "watches" the fish via live video feed and uses AI programming similar to the software embedded in Jorge Heraud's See & Spray bot to identify aberrations in color and texture on the fish's scales. Just as See & Spray learns to distinguish between weeds and crops, Sting Ray learns to differentiate between a louse and the salmon's speckled scales. When it detects a louse, the bot zaps it in milliseconds with a surgical diode laser beam, the kind used for eye surgery and hair removal. The fish's mirrorlike scales reflect the beam, but sea lice are gelatinous, roughly the consistency of an egg white, so they fry to a crisp and float away.

Aarskog teamed up with two of his competitors, Lerøy Seafood Group and SalMar, to back this project with $1.5 million in seed funding. The companies first began testing the robot in 2014 and there are now about 200 devices incinerating lice at farms throughout Norway and Scotland. Still, Aarskog is only mildly impressed with the technology. "It's a mechanical version of an ancient method of lice control," he tells me. The Sting Ray mimics what so-called cleaner fish, such as wrasse and lumpsuckers, do in the wild: they nibble the lice one by one off the fish's scales. Aarskog has been filling his cages with schools of these cleaner fish for years as a control measure for lice populations, but they can't eradicate big lice outbreaks. The fish also require special foods and elaborate seaweed habitats constructed inside the cages.

John Breivik, general manager of Sting Ray, says that robotics improves upon the lumpsucker. "For every 100,000 salmon, you might need 10,000 cleaner fish to keep lice populations at bay—or only one to two robotic lasers." He stresses that the cleaner fish and the robots can work in concert. The cleaner fish are better at getting the lice that hide beneath the fish's gills, while the robots can target the colorless baby lice that the cleaner fish can't see. "It's a synergy of old and new," he observes. Sting Rays achieve about a 50 percent reduction of lice populations in the farms where they're used, and their AI systems grow smarter and more effective at targeting the lice over time. "The effect is like compound interest," Brevik says. For his part, Aarskog is less sanguine. "We'll see," he says flatly, peering into the icy water of his fish cage as it flickers with beams of light.

Having spent years testing new approaches without success, Aarskog is understandably wary. About a decade ago, when the lice problem first started getting out of hand, he and other industry leaders were lacing their fish feed with emamectin benzoate, a chemical marketed as

A Sting Ray bot gets lowered into a fish cage

Slice that passes through the lining of a fish's gut and into its tissues, where the sea lice absorb it and die. The chemical worked for a while, until the sea lice developed resistance. Aarskog and others tried hydrogen peroxide baths, rinsing their fish in the chemical every few weeks as they matured. Again the lice adapted. They tried hydroblasting—running lice-infested fish through a kind of aquatic car wash. This was expensive and so traumatic for the salmon that it stunted their growth.

Now Aarskog is testing other mechanical approaches in addition to the Sting Ray, including a cage large enough to hold 150,000 fish but also mobile. If the fish are threatened by a lice outbreak, the cage can be plunged deeper into colder layers of water, where lice can't survive. He's also exploring mesh "skirts" that are wrapped around cages with microscopic holes which the lice can't penetrate, along with better underwater cameras and digital sensors that could help catch lice outbreaks sooner.

If all else fails, Aarskog is also preparing to completely quarantine his fish. He's invested tens of millions of dollars in the development of spherical cages with solid polymer walls. These cages, called Eggs, are 150 feet deep and 100 feet wide, and each is capable of holding 200,000 salmon. The Eggs are a departure from the coastal Norway aesthetic—they look like white UFOs when partially submerged in fjords—but they're completely impervious to parasites. And because they promise a range of other environmental advantages, including eliminating waste, disease, and escapes, the Eggs and other so-called closed containment systems in development have gained support from environmental groups and coastal communities.

The technology, though, is costly and complex. The water in the Eggs must be pumped in from deeper layers of the ocean, continually refreshed, and filtered for microscopic contaminants; fans inside the container must produce carefully controlled currents for the fish to swim against (salmon are distance swimmers and can't build suitable muscle mass in still waters); buoy systems are needed to absorb the impact of currents in the waters outside the Eggs (clashing currents can in fact make the fish seasick); huge quantities of waste must be captured and processed; exhaustive cleaning and hygiene practices must keep the container and the fish clean and disease-free. Aarskog is also investing in a doughnut-shaped containment system (dubbed, inevitably, the Donut) which functions much like the Egg but is designed to generate a stronger, more controlled current for the salmon to swim against—producing "fitter" fish.

These closed systems are, in a sense, the aquatic equivalent of a vertical farm—controlled, hyperengineered environments. Closed containment aquaculture, in theory at least, will be able to withstand the pressures of warming oceans: water can be pumped in from ever deeper and cooler layers of the ocean, and treated for acidity. But all this will add cost. "It might seem absurd to spend this kind of money to right the wrongs of aquaculture," Ingrid Lomelde tells me, "if only global demand weren't soaring, and wild fisheries weren't in such rapid decline."

THE CHINESE FIRST began farming carp around 1000 BC during the Zhou dynasty. Long before any kind of ecosystem science existed, they intuitively developed what's now known as polyculture—an integrated system connecting aquaculture with vegetable and livestock production. Animal and fish excrement functioned as fuel for the system: manure from ducks and pigs was used to fertilize algae in aquaculture ponds and then the tiny nitrogen-rich plants were consumed by the young carp raised in the ponds. As the carp matured, they were transferred to the flooded waters of rice paddies, where the fish ate weeds, insects, and larvae that might harm the rice crop, and fertilized the crop with their own nitrogen-rich waste; the rice plants in turn protected the fish from sun and avian predators. "There's deep ecological wisdom in this symbiotic rice-fish system," Josh Goldman of Australis tells me. "More rice and fish are produced on less land than if they were farmed separately, and it cuts down on the cost of fertilizers, pesticides, and labor." Polyculture systems have stood the test of time and today are still used on millions of acres of rice paddies in China.

But most modern aquaculture in China and elsewhere follows, like Aarskog's salmon production at Marine Harvest, the monoculture agribusiness model, producing a single product on a mass scale. The monoculture approach in hatcheries presents a serious problem of inbreeding, which critics insist compromises the health of the fish population, and in the long term inhibits the success of aquaculture operations. Increasingly, new smaller-scale aquaculturalists like Goldman are trying to restore ancient principles of polyculture and apply them to large-scale fish farming. "The idea is that the waste produced by caged fish seeps into the surrounding water to nourish other farmed products," says Goldman. It's as much about managing pollutants as it is about maximizing productivity.

Goldman set up his first aquaculture farm in the early 1980s as an undergraduate at the progressive Hampshire College—he hand-built a system in which fish waste fertilized the production of kale, lettuces, tomatoes, and berries. Today, Australis is headquartered twenty minutes away from Hampshire in Turners Falls, Massachusetts, where Goldman produces about 300,000 barramundi in an old airplane hangar. He first attracted New England chefs who were able to offer the fish as "local catch," even though its natural habitat is in Australia and Southeast Asia. Soon food chains caught on—he got picked up as a supplier to Whole Foods, Blue Apron, Sizzler, and the Leonardo DiCaprio–funded company Love the Wild, which sells prepared frozen farm-raised fish in supermarkets nationwide. But as Goldman's demand has grown, he has begun to rethink the farming model that has driven his company's success.

The world has changed a lot since Goldman started out fourteen years ago: "The emergence of climate change as the overarching environmental issue of our time, along with more accurate ways to measure climate impacts, has forever altered my thinking," he says. In recent years, Goldman has gradually moved the majority of his production to Vietnam. While his product lost the "local catch" cachet, Goldman says, "it tastes better to grow the fish in Vietnam and there's a lower carbon footprint. It's counterintuitive, but farming barramundi in its natural habitat, freezing it, and shipping it to market actually uses less resources compared to the local model."

When I visit his Turners Falls facility, I can see why. The place looks to me like the aquatic equivalent of a Perdue chicken coop, with fish crammed by the thousands in each small tank. The farm in Turners Falls needs a water-treatment system big enough to filter over 50 million gallons of water a day. The systems are inherently energy intensive, as very large volumes of water need to be pumped, purified, and oxygenated. "It's like running an intensive-care unit," Goldman says, fishing a pale, dead barramundi out of a tank. "You have to control every single aspect of the fish's well-being, and they're always ten minutes away from dying. One pump or valve fails, they're gone." While ocean-based closed containment systems like the Egg are logistically complex in many of the same ways as land-based aquaculture, they have the key advantage of an unlimited and easily accessible water source.

Farming off the coast of Vietnam also means Goldman can apply some of those ancient Chinese principles of integrated aquaculture and grow other marine products alongside his barramundi. Around his fish cages, he's placing long curtains of *Asparagopsis* and other types of edible seaweed. These ancient aquatic plants store carbon dioxide and convert the nitrates and phosphorus from the fish waste to plant tissue. A key to sustainability, says Goldman, is learning to eat foods that occur in the beginning of the food chain, and seaweed is almost as incipient as it gets. He plans to sell the seaweed fresh to Asian and Polynesian markets for salads and soups, and to dry and grind it up for cattle feed as an alternative to corn-based feeds in the United States. Bren Smith, another Massachusetts-based aquaculture entrepreneur, has developed a similar polyculture model for his start-up, Greenwave. Off the coast of Cape Cod he's growing huge expanses of kelp on a 3-D grid, along with cockles, mussels, and scallops that feed on fish waste. In cages below are oysters and clams that consume the heavier organic waste that falls to the seafloor.

Alf-Helge Aarskog isn't practicing polyculture per se, but the Egg will make it possible for him to capture all the fish waste, harvest it, and produce two new revenue streams—biofuels and high-quality fertilizers. He's also testing a new feed for his salmon that will draw ingredients from a source even lower on the food chain than seaweed. Down the line, he says, the most critical issue for sustainable aquaculture may not be where or even how the fish are farmed, but what they're fed.

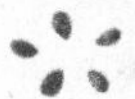

MONGSTAD IS ANOTHER village tucked in the elbow of a Norwegian fjord, a few hundred miles south of Frøya, but this place looks more like Newark, New Jersey, than an Old World fishing town. It's the center of Norway's oil-refining industry—which is, like the country's oil supply, on the decline. It's also the location of the strangest, if not the most important, Marine Harvest laboratory.

The lab, which sits adjacent to a vast snarl of industrial steel pipes and blinking lights, is housed in a gleaming glass structure that's completely transparent—a brightly lit greenhouse. No flowers or vegetables are growing inside, instead there are clusters of glass cylinders that look like giant test tubes stacked horizontally, each about eight inches wide by forty feet long and filled with liquid. The liquids are varying shades

The Marine Harvest algae lab in Mongstad

of green—emerald, viridian, forest, and a muddy greenish black, some thin and watery, others nearly opaque. Lab-coated scientists are peering at them, taking notes, turning dials on the racks that hold the tubes.

Six strains of phototrophic algae with names like *Nannochloropsis gaditana, Tetraselmis chui,* and *Phaeodactylum tricornutum* are growing inside the tubes. The lab is located next to the oil refinery because it's siphoning carbon dioxide, a key ingredient that the algae need to grow. CO_2 is captured in the oil-refining process and piped into the laboratory, bubbling into the glass tubes. The CO_2 combined with sunlight fuels the algae growth. Algae are, you might say, the OG of the aquatic food chain. Aarskog believes these tiny microscopic plants that, millions of years ago, were among the first living organisms to grow on planet earth now hold a key to sustaining aquaculture.

Øyvind Oaland, Aarskog's global director for research and development, oversees this laboratory. He's a lean, balding man with a tidy gray goatee and thick-framed glasses. Among his many R&D challenges, the one that fascinates him most, he says, is devising strategies to convert carnivorous fish to a vegetarian diet. "Salmon are fish eaters, but they don't require fish per se to succeed," he tells me. "They require the nutrients and fats in those fish. If we can derive those same nutrients and fats entirely from plant sources, that changes the game."

Vegetarian salmon feed would liberate aquaculture from dependence on wild fisheries, which are prone to collapse. Fish farmers have been getting much more efficient in the amount of fish they put into salmon feed. As a teenager, when Aarskog was dumping sardines and herring into the wooden tanks, the salmon ate as much as six times their weight in wild fish. In the 1980s, salmon farmers began to transition to manufactured feed pellets, which meant that marine ingredients in the fish feed could be mixed with plant-based ingredients. This brought down operating costs and fueled the growth of aquaculture. The pellets were less prone to spoiling, could be stored over long periods, and were easier to distribute to large cages of fish.

Marine Harvest's fish feed division is now responsible for more than a third of the company's total revenue. The pellets, which are fine for human consumption (I ate one and it tasted like a moldy sneaker, but nothing bad happened), are made with about 75 percent terrestrial grains (corn, wheat, soy), 20 percent fish meal (ground-up fish, mostly from the heads, tails, and bones that are cut from packaged fillets), and 5 percent fish oils. Despite consumer concerns about fish eating terrestrial grains (what self-respecting salmon would eat *corn*?), aquatic veterinarians say that biologically it makes no difference to a salmon if it gets its proteins from plants or other fish. Environmentalists applaud this shift because it curbs the slaughter of wild fish. Consumers like me also appreciate it because vegetarian fish feed brings down the cost of farmed salmon.

Two decades ago, the industry consumed about five pounds of wild fish for every pound of salmon it produced. Today that ratio is down to 0.7 pounds of wild fish for every pound of salmon, says Aarskog. It's progress, but still a lot of wild biomass in an industry that's seen meteoric growth. The problem going forward isn't finding substitutes for fish meal—that can be produced from plant proteins and even ground-up insects. The challenge is finding fish-oil alternatives. Technically you could feed salmon a vegetarian diet without the fish oil and they'd grow just fine, but their meat would contain no omega-3 fatty acids. "It would be basically like pink chicken," Aarskog's communication director, Ola Hjetland, tells me. "You lose the nutritional value that makes this a prized product."

Here's where the algae come in. Algae manufacture the omega-3 fatty acids that are then passed up the food chain to krill and other crustaceans, then to the small fish that eat the krill, and eventually accumulate in the salmon. "By manufacturing omega-3s directly from microalgae, we're moving back to the origin of the nutrients," says Oaland. Omega-3s manufactured from algae are already added in small amounts to milk, eggs, and orange juice—the process has been proven; it's a matter of finding ways to achieve a large increase in scale.

Some critics balk at the prospect of vegetarian salmon feed ("You can't feed lentils to lions!" Don Staniford tells me), but Oaland insists we have to shed old ways of thinking about sustainability. "We're moving toward a future with much more demand and many more environmental constraints on fish production. We have to ask the question 'How do we produce food out of non-food? How can we become radically more creative?' " He sees algae as a non-food that has the potential to become, like seaweed, a far bigger part of the modern diet, if not as direct sustenance, then as a fuel to grow good food.

Salmon, he concedes, are less resource-efficient than other fish. They consume more food than do tilapia and catfish and cod, for example. But compared to land animals, they're frugal eaters. Fish need fewer calories because they're cold-blooded and they don't have to partition energy to heat their bodies or build layers of fat and fur for warmth. They live in a buoyant environment and don't have to resist gravity or walk upright on four legs. "It takes roughly a pound of feed to produce a pound of farmed salmon; it takes almost two pounds of feed to produce a pound of chicken, three for a pound of pork, and about seven for a pound of beef," writes *National Geographic* reporter Joel Bourne. "As a source of animal protein that can meet the needs of nine billion people with the least demand on the earth's resources, aquaculture . . . looks like a good bet."

More than once in our conversations, Aarskog stressed that a far greater challenge than making aquaculture sustainable will be persuading people to eat more fish and less meat. While seafood is the dominant source of protein for nearly half of humanity, relatively little of it is eaten in the industrial West, particularly in the beef-obsessed United States. Here, the environmental problems that come with raising cattle far surpass those that come with sea lice and lagoons of fish poop. Solving these challenges will require, if not huge increases in the cost of beef, then innovation stranger and perhaps more unsettling than robotic lice lasers and phototrophic algae farms. It may require rethinking beef altogether.

CHAPTER 8

Meat Hooked

The human mind prefers something which it can recognize to something for which it has no name . . . yet a small microscope will reveal wonders a thousand times more thrilling than anything Alice saw behind the looking glass.

—DAVID FAIRCHILD

MEAT IS THE old sweetheart I can't quit. Having taken to heart the case against it, and the lessons of the previous chapters, I know it's a toxic entanglement. I've tried several times to phase meat out of my diet completely, just as I've phased out sugar, coffee, and alcohol for long periods (mostly without missing them). But with this particular vice, I'm weak-willed, and living as I do in Nashville, home of BBQ and hot chicken, I'm a shark in chummed waters. I keep circling for more—and also for justification for why it's so hard for me to give it up, trying to uncover a way to keep some meat protein in my diet without hastening my own (and the planet's) demise.

A doctor friend I consulted about my meat cravings told me that it may be in my blood: "You're a type O carnivore." A yoga teacher told me that it's a matter of my *dosha*—the energies that are believed in Ayurvedic medicine to govern physiological activities. She explains, "You're *vata*—you need dense, grounding, protein-rich foods." Whether or not these theories are true, I'm pretty sure nostalgia plays a role. I was raised in the 1980s on a rib-sticking diet so centered on red meat that chicken and fish were regarded as vegetable matter. And while seafood remains the dominant protein source for people globally, millions more, it seems, would like to be raised as I was: the processing and consumption of beef, pork, and chicken has nearly doubled worldwide in three decades, and it's expected to double again by 2050.

Beef is the real killer among these. Over the years, I've come to understand that my red meat habit was draining America's lakes and rivers, increasing my risk for heart disease, contributing to the destruction of virgin rain forest cleared for cattle grazing, and driving global warming. Livestock production accounts for about 15 percent of all greenhouse gas emissions globally, more than all forms of transportation combined. It also gnaws at my conscience that most animals raised for slaughter aren't given decent living conditions.

My undergrad students, at least a third of whom are some version of vegetarian, have made the case that even with occasional meat consumption I'm endorsing animal cruelty; they've shown me hidden-cam videos taken inside cattle slaughterhouses and of caged pigs and abused chickens. They've argued convincingly that most of us way over-consume protein (the average American adult eats about 100 grams of protein a day, but a healthy adult needs only about 50 grams) and, on top of all else, meat is dirty. "Do you have any idea how much cow shit is in minced beef?" one young vegan challenged. She showed me a *Consumer Reports* investigation in which more than three hundred packages of hamburger meat were tested for fecal contamination, and every one was befouled—the maraschino cherry on top of my meat-shame sundae.

So when my brother, a climate scientist raised in the same meat-centric household I was, vowed after his own protracted struggle at the age of forty-seven to cut beef from his diet, I agreed to do the same. I embraced nut butters and tofu, swapped out bolognese for ratatouille, and gave up tender brisket for veggie burgers, all of which "come in two flavors of unappealing," as one reluctant vegetarian groused in *Outside* magazine, "the brown-rice, high-carb, nap-inducing mush bomb, and the colon-wrecking gluten chew puck." I succeeded for sixty-four days—until someone offered me a plate of carne asada tacos with pickled onions, jicama slaw, and a to-die-for creamy dill sauce. I devoured them like a starved hyena.

And so I found myself confronted again with my dilemma—and the world's. How do we fix the overwhelming problems of meat production when informed, penitent eaters like me can't curb their bloodlust? The question led me in many directions: First, to the farm of Sam Kennedy, a thiry-five-year-old former marine who's raising cattle and sheep on 3000 acres in central Tennessee using a practice championed by regenerative farmers known as "managed grazing." Kennedy is a devotee of Joel Salatin, the livestock farmer who appears in Michael Pollan's *The Omnivore's Dilemma,* and of Gabe Brown, who practices managed grazing on a 5000-acre farm in North Dakota. Kennedy converted sprawling tracts of land that had once grown corn for animal feed into pastures with tall, native grasses that function as rich carbon sinks. Kennedy's cattle and sheep munch the wild grasses while pounding the land with their hooves, fertilizing the soil as they mix in their manure, causing new grasses to grow.

The practice mimics what herds of wild herbivores have been doing for millennia, only Kennedy uses contemporary tools like drones and movable electric fencing to rotate his herds through sections of pasture, creating a continuous cycle of regeneration. This is third way thinking at its best, elevating traditions with modern techniques. But while Kennedy's farm is storybook beautiful, his animals are well-treated, and his meats are bar-none delicious, his products are also expensive, at or above Whole Foods prices. They're worth the effort and expense, but Kennedy cautions that, for all its promise, managed grazing is likely not going to replace mass-scale beef production.

"It boils down to scale, not just on the farming end, but on the processing end," he says. "It costs Tyson $125 a head to get beef slaughtered and processed; it costs a local farmer like me $550 a head." Kennedy's approach is a promising and climate-smart way to produce high-quality meats going forward, but I'm left wondering about consumers on a tight budget: Can sustainable meats be affordable for the masses? Again, I cast about for more answers—in the wrong direction, at first.

Beef producers in China, I'd read, were cloning more than 100,000 head of cattle in an effort to satisfy the country's rocketing meat demands. I tried to get permission to visit this operation to find out why, but it wasn't granted. So I sought out Ty Lawrence, an academic researcher who in 2016 cloned the first steer on American soil. By 2018 he had cloned three heifers and bred dozens of calves from cloned parents for his research on efficient, high-yield beef production. Lawrence's title includes "Director of the Beef Carcass Research Center at West Texas A&M" and he lives, per his online bio, in the Amarillo suburbs with his wife, kids, and dog, Fuzz.

I hopped a plane to northern Texas. There I met Lawrence and his veterinarian, Gregg Veneklasen, a practitioner of animal cloning. Veneklasen had a colorful history that involved cloning prized racehorses for famous magnates, as well as antlered bucks and exotic game for private hunting grounds. Now, he said, he was shifting his focus from "cloning for entertainment to cloning to feed the world."

Cloning cattle presents a huge potential cost advantage because identical carcasses could allow beef slaughter to become fully automated (some of it is currently done by hand, carcass by carcass, because there's so much variation in the way cows amass muscle). The environmental advantages came down to less fat, Lawrence said. He was cloning only steers with thin fat layers, which meant they converted minimal energy to fat, and could reduce the total feed and water required to produce prime beef by about 5 percent.

Alpha, America's first cloned steer

Setting aside the ethics of animal cloning, 5 percent better efficiency in beef production is good—but it's also not enough progress in an industry that's poised to double in the near future. The average steer, lean or not, weighs about 1000 pounds and produces only about half its weight in meat. The other half of the animal (bones, hide, viscera) also require tons of feed and water to produce, yet they're inedible. It means, in short, that about half of all the corn, water, and greenhouse gas emissions associated with beef production go to making animal parts that don't get eaten—a major inefficiency in the food system. This is one reason that several new and old players in the meat industry are exploring methods of chicken, pork, and beef production that don't involve animals at all. So I changed my track, traveling 1400 miles northwest to the Bay Area, a giant cultural leap from the Texas badlands.

THE LABORATORIES OF Memphis Meats are located in Berkeley, California, next to a juice bar and a small-batch coffee roaster, and not far from Alice Waters's Chez Panisse, the Shangri-la of farm-to-table restaurants. Here, in a newly renovated brick building on a quiet, tree-lined street, a group of scientists is rethinking meat production at the molecular level—which is to say, completely. Their novel approach both upholds and upends everything that the patron saints of Berkeley's sustainable food movement hold dear.

Cofounded in 2015 by Uma Valeti, an Indian-born cardiologist, and Nicholas Genovese, a stem cell biologist, Memphis Meats is the world's first start-up to grow meat in a laboratory using tiny samples of muscle, fat, and connective tissues taken from living animals. "We are a meat, poultry, and seafood company that makes end products no different than conventional meat, while eliminating the need for animal slaughter," Valeti tells me in a phone call before my visit. He adds that the cells that are grown, or "cultured," in his laboratories are "alive," even though they're not attached to the animal. They're so alive, in fact, that the mature muscle tissue he produces actually responds—as in flexes, or spasms—when stimulated.

The notion that a serving of cultured meat had once been flexing in a petri dish would send me running to the tofu section, I tell Valeti. But he goes on to outline the many benefits that might coax me right back: "Cultured meats are identical on a cellular level to animal meats and can be as or more nutritious and delicious," he says. Moreover, the production process could reduce the greenhouse gas emissions from meat production by more than three-quarters, while also cutting associated water use by up to 90 percent. Cultured meats could also eliminate the risk of bacterial contamination (gone would be the threat of *E. coli* and the helping of feces) and reduce the risk of heart disease and obesity (fats and cholesterol levels in these meats can be controlled). "We're talking about changing the lives of billions of humans and trillions of animals," Valeti tells me.

I first heard about Memphis Meats in early 2018 when Tyson Foods announced plans to invest in Valeti's start-up. The investment sounded preposterous to me, coming as it was from a company that produces one of every five pounds of meat consumed in the United States. Annually, Tyson sells $15 billion worth of beef, $11 billion of chicken, $5 billion of pork, and $8 billion in prepared foods under a brand roster that includes Hillshire Farm, Jimmy Dean, and Ball Park Franks. About half of Tyson's fresh and frozen meats are sold in fast-food restaurants, including McDonald's, Burger King, Wendy's, and KFC. Why would an industrial meat company be backing the production of such an obscure, even Frankensteinian, product?

At the time, Tyson's CEO, Tom Hayes, had vowed to transform his eighty-three-year-old business into "a modern food company." He trumpeted the promise of "sustainable proteins" and "zero-carbon foods." He made statements like "I took this job to revolutionize the global food system" and pledged "to raise the world's expectations for the good we can do through food." These promises sounded dubious coming from a man whose company processes about 1.8 billion animals a year and is responsible for more greenhouse gases than the whole of Ireland. Yet Hayes insisted that it's precisely because of its scope that Tyson has the potential to make a global difference. "We're so big that the industry can't change if we don't lead," Hayes told me.

Hayes was investing not just in lab meats but also in start-ups making plant-based proteins—most notably, Beyond Meat, which produces burgers, sausages, and nuggets made from soy and pea proteins, products that are now sold in more than twenty thousand grocery stores. The broader category of "alternative meat" products in the United States has been soaring in recent years. The Silicon Valley start-up Impossible Foods has raised more than $680 million to push its product—plant-based hamburger meat flavored with synthetic animal blood—into the mainstream. Recent Nielsen data shows that in a one-year period, retail sales of meat alternatives jumped more than 30 percent—many times the growth of meat sales and of retail food sales generally. Another study found that 70 percent of meat eaters are substituting nonanimal proteins at least once a week.

"If you can't beat 'em, join 'em, right?" said Hayes. He pointed to the ongoing disruption of the automobile industry by electric-car technology and of the tobacco industry by vape technology. He purported to welcome this kind of change. "We want to actively disrupt ourselves—we don't want to be disrupted. We don't want to be Kodak."

Tyson Foods isn't the only player in the conventional meat industry making unconventional investments. Cargill Meats, another one of the world's biggest producers of beef and poultry, had invested in Memphis a few months before Tyson jumped in. Valeti's company has also drawn tens of millions of dollars of investment from more predictable players: Bill Gates, Richard Branson, and the venture firms Atomico and DFJ, which focus on disruptive technologies. Cargill's Sonya Roberts described Memphis Meats as developing simply "another way to harvest meat. For people who want a product from an animal welfare perspective, we want this to be there for them."

Backers of lab meats are betting that plant-based products will never achieve the depth of flavor or the full "mouthfeel satisfaction," as Valeti calls it—the bounce, chew, and density—of meat. "Faux hamburger patties and nuggets are an important application, but a narrow one," he says. "Humans have evolved over millennia eating animal meat. More than 90 percent of the global population eats it today. What they want is a product that tastes, handles, and cooks exactly like traditional meat."

What Uma Valeti's trying to produce, in other words, is not an *alternative* to meat, but a whole-hog (or whole-cow) replacement—without the bones, organs, hide, "oink," or "moo." It's a wildly ambitious goal, however strange and unsettling. Which is why, when Valeti invites me to visit his laboratory and taste a sample of his product, I accept.

"KEEP IN MIND that what you're about to eat came from a cell, not a slaughterhouse," Valeti intones. "I just want to be sure you're looking at this as a very big, historic thing."

So begins the dramatic preamble to what is certainly the most expensive and possibly the most precedent-setting meal I'll ever eat—a 2-ounce, several-hundred-dollar portion of duck breast grown from a cluster of cells that were taken from a live duck presumably still waddling around a farm in Petaluma, California. I hesitate only a little before I sign a legal waiver acknowledging that cultured meats "are experimental, their properties are not completely known," and agreeing to "voluntarily accept all risk of loss, damages, injury or death that may occur to me as a result of my participation in this tasting."

Uma, who has mission impossible hair and a hundred-watt smile, leads me and half a dozen members of his team into the gleaming white-and-stainless-steel test kitchen adjacent to the laboratories in Memphis Meats' headquarters, which have been newly built to accommodate the quadrupling of its staff in the past year, from nine employees to thirty-six. Together we watch a small, pale lump of duckless duck meat sizzle in a shallow frying pan. Valeti explains that he's working on duck, in addition to the more common American meats, because it's highly popular in China, where meat demand is surging.

Uma Valeti

"Notice the way the meat smells as it sears—you can't get that rich aroma from plant-based products," observes Valeti. He has instructed the in-house chef to use a neutral oil and season the duck breast with nothing but salt and pepper so that I can experience the meat's unalloyed flavor.

The chef places the golden-browned specimen next to a colorful bed of radicchio, cabbage, orange slices, and fresh figs dressed in a citrus vinaigrette and invites me to sit at a table set for one. Uma and his cadre remain standing and watch me expectantly. "Should I say grace or . . . give thanks for the cell donor?" I joke, feeling antsy under the pressure. "In my house we just say, 'Rub-a-dub-dub, thanks for the grub,'" says Eric Schulze, a senior member of the Memphis team. I repeat the line, lifting my fork and knife and angling the blade. "Try picking it up in your hands and playing with it first," Valeti interjects. "Pull it apart, feel the crispiness and density, see how it falls apart." I oblige, prying at the knob of meat. It's taut and springy and feels a bit like trying to split a bouncy ball. But as the meat comes apart I see what Valeti's talking about—the long, stringy muscle fibers bind and hold together, stretching and then separating as I pull. "A far cry from a garden burger," I say. Valeti nods enthusiastically.

I pop a piece in my mouth and the duck tastes as advertised—meaty.

I've eaten duck only a few times in my life, but I know it tends to be chewier and fattier than chicken. This duck strikes me as a bit *too* chewy (I have to put my jaw into it) and overly stringy, with a vaguely metallic aftertaste, but it's certainly familiar-tasting, and I have no problem eating every bite. It's clear to me that if I ate the stuff gussied up with sauce and trimmings, as Peking duck or duck à l'orange, I would be hard-pressed to distinguish the meat from conventional duck. And it's ultimately this—the familiarity, the passability, the very *ordinariness* of the product—that's so remarkable, given its extraordinary provenance.

UMA VALETI'S PATH toward Memphis Meats began when he was twelve years old, attending the birthday party of a neighbor in Vijayawada, the city in southern India where he was raised. Musicians performed and guests danced as the host family served steaming platters of curried goat and chicken tandoori to those gathered in the front yard. At one point, Valeti wandered to the back of the house, where he found a bloodbath: the cooks were decapitating, gutting, and cleaning the chickens that would be added to the next batch of tandoori. "It hit me all at once that there was a birthday celebration in the front of the house, and a death day in the back. It was enormous happiness and enormous sadness existing in the same moment."

While roughly a quarter of Indians practice vegetarianism, Valeti's family did not. On Sundays, his father, a veterinarian, would bring home half a kilo of goat or chicken, which was all the meat the family of four would eat for the week. "I loved Sundays because the smells in the house would become really delicious," Valeti recounts. He continued to eat meat for years after the "death day" experience. His dad took care of livestock, mostly cows and sheep, and when Uma tagged along with him on farm visits they often discussed human-animal relationships.

Valeti's biggest influence was his maternal grandfather, who lived with his family and worked as a general physician. He'd served in the Indian independence movement led by Mahatma Gandhi and marched alongside the freedom fighters. "Tatayya was a complete believer in self-reliance. He wore only clothes that he spun and sewed himself," Valeti tells me. "He practiced medicine for free, following the belief in intentional service." Tatayya's wife died when giving birth to their fifth child, so he raised Valeti's mother and her siblings alone. They lived on the vegetables and grains they grew on their land and the offerings they received from the patients Tatayya treated.

Valeti followed his grandfather into medicine, entering medical school at age seventeen at the Jawaharlal Institute in Pondicherry, with a focus on cardiology. In pursuit of a visa that would enable him to study in the United States, Valeti accepted a residency in Jamaica. There he met his wife, a pediatric ophthalmologist; they both went on to residencies at the State University of New York in Buffalo. Valeti moved on to the Mayo Clinic in Minnesota, where he became a leading researcher in interventional cardiology while diagnosing and treating heart disease.

"I arrived in the United States to see the KFCs, McDonald's, and Pizza Huts everywhere," Valeti recounts. "I walked into supermarkets and was fixated on the massive meat cases." He loved the flavors, "especially the golden fried chicken," and was mesmerized by the shrink-wrapped packages of links, chops, breasts, drumsticks, cutlets, mince, and T-bones. "The presentation was amazing to me," he says. But the scale of it was terrifying. Valeti dove into research. He found that tens of billions of animals are raised annually to feed 7 billion people worldwide; that the United Nations expects global meat consumption to grow from 250 billion pounds today to nearly 500 billion by midcentury; that livestock production gives off more greenhouse gases than all transportation combined; that cardiovascular disease, the number one killer in the world, highly correlates with excessive meat consumption.

Valeti stopped eating meat—a decision that felt like a paltry response to such a pervasive problem. Somewhere around this time he also stumbled onto an article written in 1931 by Winston Churchill predicting future trends: "We shall escape the absurdity of growing a whole chicken in order to eat the breast or wing, by growing these parts separately under a suitable medium."

Valeti's breakthrough idea came in the early 2000s, when researchers were beginning to use stem cells to regrow everything from bladder linings to brain tissue. Stem cells are renewable and self-replicating, and they can become different types of tissue as they mature. As part of a clinical study, Valeti was injecting stem cells into patients' hearts with the hope that they would replace and regrow damaged tissue. It struck him that if human tissues could be grown for medical purposes, why couldn't animal tissues be grown for food?

Valeti wasn't the first to have the idea—in fact, the basic technology for growing muscle tissue had been around for two decades, but the only work being done on cultured meats was in academia. He wrote letters to Dutch scientists who were pioneering the research, but didn't hear back, so he established a lab of his own at the University of Minnesota. Here, he teamed up with Nicholas Genovese to try to replicate chicken breast tissues. By 2015, they'd proven it was possible. When Valeti sent his results to the venture firm Indie Bio, he heard back within an hour. They wanted in, on the condition that Genovese and Valeti would move their new enterprise to San Francisco.

For Valeti, this meant ending the cardiology work he loved, living apart from his young kids and wife, who led a private practice in Minneapolis, and commuting home on weekends. But he knew that for the idea to succeed he had to get out of academia, so the Memphis Meats founders struck out for the Silicon Valley suburbs. They named the company not for the meat-loving city in Tennessee, but for Memphis, Egypt, an ancient center of innovation. Raising more money for their endeavor was "harrowing," says Uma. "We got flat-out rejected by thirty to fifty investors." But by 2016, when Valeti and Genovese debuted the world's first lab-grown beef meatballs, costing a mere $18,000 per pound to produce, investors started coming in droves.

Memphis Meats does fried chicken

The products Valeti and his team are now developing—described variously as "cell-based," "cultured," "clean," and "in vitro" meats—are expensive to produce and still years from hitting the market. But the cost of production has come down—from about $1 million per pound in 2015 to a few thousand dollars a pound today. "We've got lots of work to do bringing down cost, improving flavors and textures," Uma tells me, "but the improvements we'll make in the next three, five, ten years will be immense."

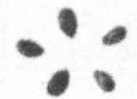

MEMPHIS MEATS' BERKELEY headquarters are bright and elegant, with dove-gray carpets, huge windows, and open-plan seating. Hanging in the lounge area above the vegan-leather poufs is a mosaic of album covers that includes employees' favorite food-themed bands: Meat Loaf, Red Hot Chili Peppers, Salt-N-Pepa, Counting "Cows."

There are four Memphis Meats laboratories, and while all look like typical biology labs with their microscopes, centrifuges, eyewash fountains, and racks of beakers, Valeti insists on calling them "farms." "Growing cells is somewhat analogous to growing animals," he explains, "so our process flows in much the same way as a farm's does." The first lab is dedicated to what Valeti calls the "cell line development team." These folks select the best breeds of cells from which the meat is grown. Genovese has developed partnerships with farmers in California and beyond who grow various types of livestock, from heritage animals to breeds that have been selected to yield a higher quality of muscle or certain flavor profiles in their meat. The partners send Genovese small amounts of tissue from the part of the animal he wants to replicate—no more than is collected in a typical biopsy.

In theory, any part of an animal can be grown, including bones and organ tissue, but for now Valeti's team is growing only what we directly eat—muscles, connective tissues, and fats. The cells are stored in liquid nitrogen and then "reanimated" from the deep-freeze state. From these samples, Genovese and his team can identify and select the healthiest cells that are most capable of renewing themselves, and therefore easiest to grow.

During our tour, Valeti pulls a petri dish from an incubator, places it under a microscope, illuminates the base, and invites me to look in. "See the tiny wormlike things that look like elongated triangles? Those are muscle-forming cells—or 'starter cells,'" he says. I adjust the scope, focusing in on what looks to me like a scattering of pale, glinting stars at twilight. I begin to notice that two cells are round, rather than oblong, and nestled next to each other in a couplet. "A cell dividing!" cries Uma.

It's the first time I've witnessed this essential miracle of life—that living cells renew themselves. They can, in theory, replicate themselves indefinitely, but they need the right conditions and fuel to do so. Which leads us to the next laboratory, occupied by the "feed development team"—the scientists who formulate the special brew of nutrients in which the cells are grown. K. C. Carswell, a Memphis Meats scientist, explains the complexities of feeding cells: "A cell can't eat a whole blade of grass—it eats the subcomponents of that grass, which in the body of a cow are broken down in the stomach." Carswell describes her work as a form of "biomimicry"—an effort to simulate processes that exist in nature. "We're trying to give cells access to the same nutrients and growth factors that they would get within the cow."

Carswell's team tests dozens of configurations of feed daily, each a different mixture of proteins, fats, hormones, carbohydrates, vitamins, and minerals suspended in water. This broth serves as a substitute for blood, which delivers the nutrients consumed by an animal into its cells. Historically, scientists growing cultured tissues have used fetal bovine serum, a substance extremely rich in the nutrients that cells need to grow, but there's a serious hitch—it's extracted from cow fetuses, which makes it financially, environmentally, and ethically costly. Carswell and her team have been working feverishly to produce a version of the growth serum that contains no animal-derived ingredients. They've succeeded, but still need to figure out how to produce it affordably and at a high volume.

Once the cells are selected, they're placed inside bioreactors, which are essentially ultrasophisticated Crock-Pots, where they're fed the special brew. A pump continuously circulates the feed and oxygen throughout the morass of cells (there are billions within a single cubic centimeter). As the cells mature, the feed is changed according to their different stages of growth. Young cells need special nutrients as they replicate. Hour after hour, the cells elongate while growing closer together. The mature muscle cells form long chains, linking end to end while also rubbing shoulders and binding together, building layer upon layer. The chains and layers begin to look like the roiling ocean in a Japanese wood-block print, or, as Valeti describes it, "the whorls in a Van Gogh painting." Once the cells are mature, they simply need to bulk up, so the feed shifts to a simpler formula of proteins and fats. The process is similar to that of a cattle feedlot system: calves need special nutritive feed to grow and mature, but eventually they end up in a fattening lot where they eat a calorie-rich diet to put on weight.

When it's time for harvest, the solution used to feed the cells is drained from the bioreactor and the cultured cells are removed in the form of what Valeti describes as "an integrated piece of meat." The end result is not, in other words, a mushy mash of cells, but rather a solid structure with layers of fused tissues, similar to what you'd harvest from a slaughtered animal. The lab meat is technically still "alive" when harvested, even if it lacks sentience, but it's immediately stored in a freezer and "dies" in the process. "The point when the tissues are no longer considered living is the point of asphyxiation, when the cells no longer receive oxygen," says Valeti.

He knows that I'm eager to see the signs of life exhibited by the disembodied animal cells before harvest. He opens up his laptop and runs a video of lab-grown beef tissue in a petri dish that Genovese captured via a camera mounted on a microscope. "The contractions can occur spontaneously or in response to stimuli—like an electrical pulse," Valeti tells me, adding that the petri dish can also be spiked with caffeine "to get the cells fired up."

The black-and-white video rolls. I see what looks like a smear of beef carpaccio on the bottom of a petri dish. It's perfectly still—and then

Valeti shows me a magnified image of maturing muscle cells

it spasms. The strands of muscle look like tiny rubber bands getting pulled and released. I'd been expecting it, but I audibly gasp. The experience doesn't repulse or scare me, though. I feel less like Victor Frankenstein witnessing the monster's first twitch than like Alice passing through the looking glass. I'm in awe of the power of science, having entered a realm of possibility that I didn't know existed.

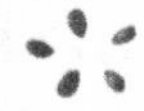

SINCE THE FOUNDING of Memphis Meats in 2015, dozens of start-ups worldwide have begun to develop lab-grown beef, pork, poultry, and seafood. Among them, Mosa Meat, Just Meats, New Age Meats, and the Israeli company Future Meat Technologies have announced plans to release cultured meat products—from chicken cutlets to breakfast sausage—in the near future, some by 2020. Another Silicon Valley start-up, Finless Foods, has been developing cultured bluefin tuna. The founders, a pair of twenty-something biochemists, say they're on track to debut their first product, a cell-based fish paste, by late 2019. SuperMeat is formulating a special animal-free brew to replace fetal bovine serum, which it hopes will fuel the rise of this young industry. Valeti won't disclose a release date for his products, nor will he specify which products he plans to release first, but says he isn't concerned that others may beat him to the punch because he's going for quality rather than speed. "It's fantastic—nothing proves the viability of an idea and shows the promise of a market better than competition," he says.

Just a couple blocks down the street from the Memphis Meats' headquarters in Berkeley are the offices of Perfect Day, a start-up that's well on its way to defining another new frontier in animal-free products. Instead of "cell-based" foods, they're creating dairy products made with "flora-based" proteins using a controlled fermentation process. Perfect Day has developed GMO yeasts that are designed to function like miniature protein factories, cranking out amino acids identical to those found in dairy milk. These proteins are the core ingredients for products that taste indistinguishable from cow's milk, and could produce the full gamut of cheeses, yogurts, and ice creams. I swung by Perfect Day during my Berkeley visit to check out their labs and sample their vegan chocolate ice cream. The stuff looked, tasted, and melted exactly like a scoop of dairy-based Häagen Dazs. Similarly, another nearby biotech start-up, Clara Foods, is creating flora-based eggs, using GMO yeasts that produce proteins identical to those in egg whites.

It's anybody's guess which of these products will succeed on a grand scale and when, but there's no question that research and investment into animal-free products is on the rise. Perfect Day, for example, has partnered with Archer Daniels Midland to begin using its fermentation equipment to produce nonanimal dairy goods at scale. So even if lab meats and fermented milk and eggs sound weird now, they probably won't by the time these products start hitting store shelves. The recent success of plant-based meat products is also helping to pave the way for mainstream acceptance. "This shift away from animal meats is already happening—whether you're aware of it or not," says Matthew Prescott, food policy director for the Humane Society. Prescott calls plant-based meats "gateway products" for the broader industry of meat alternatives.

Take, for example, the implausible success of Impossible Foods, which is headquartered thirty miles south of Memphis Meats in

Redwood City. When, in 2014, CEO Patrick Brown introduced the first version of his Impossible Burger, a plant-based patty flavored with synthesized animal blood, the media was intrigued, if skeptical. "Meet the Fake Burger That 'Bleeds,'" read a headline in the *Wall Street Journal*. The concept sounds almost as creepy as lab meats, yet the burger has already become a staple product at fast-food chains from Shake Shack to White Castle.

The linchpin ingredient in Impossible's product is "heme," short for hemoglobin—the iron-rich substance that gives blood its deep color and earthy-metallic flavor. (When I spooned a bit of heme into my mouth at the Impossible Foods test lab, it tasted oddly, eerily, like sucking on a paper cut. I was struck with the discomfiting realization that it's this—the flavor of mammalian blood—that I and other meat eaters crave in a burger.) Brown discovered that hemoglobin can be manufactured not just by animals but by the root nodules of soybeans. Using a flora-based fermentation process similar to that of Perfect Day, Brown took a snippet of the soybean genome that codes for heme production and injected it into yeast cells, which then serve as little heme factories. They produce a frothy pink liquid that gets concentrated into a thicker, deeper-hued blood-flavored serum. The rest of the burger is made from vegetable products, including textured wheat proteins, flavorless coconut oil, salt, potato protein, and other additives found in processed foods. It's not exactly a wholesome superfood, but it's a convincing alternative to minced beef. Brown says, pound for pound, his product generates less than an eighth of the greenhouse gasses created by conventional beef production. He has big plans for industry disruption: "Our mission is to completely replace animals in the food system by 2035, which we will certainly do."

In 2016, David Chang was the first to put the Impossible Burger on a menu, at his Momofuku restaurants. By 2019, the plant-based product had made its way into Burger Kings worldwide and into niche fast-food chains like White Castle, where it was selling for $1.99 and had been heralded by critics as both the best and worst burger in America. I've had three Impossible Burgers—first at the Paris Climate Conference, where burger samples were passed around as snacks, second at

Momofuku Nishi in New York City, and a third in Brown's laboratory. Each was an updated formula and better-tasting than the last, and all were a great improvement on the mushy, mealy things that have until now dominated the plant-based burger market. The Impossible Burger was hardly perfect—a bit squishy and too salty-metallic for my palate, but smothered as it was in ketchup and mayonnaise and American cheese, I ate every morsel.

It's a safe bet that Brown will eventually tweak his Impossible formula to something indistinguishable from, if not superior in flavor to, a Big Mac patty. And whether or not he succeeds, he'll perhaps have produced something of even greater value: research into the many reasons humans crave meat. The large, glass-walled Impossible laboratories at the edge of the Stanford University campus hold an array of machines, including several known as gas chromatography mass spectrometers, which analyze food chemistry down to the molecule. "We can take foods that activate the pleasure centers of your brain, a barbecued spare rib or roast turkey or apple pie, and get the data behind why it's biochemically satisfying," says Celeste Holz-Schlesinger, the director of research at Impossible. "It gives us the molecular components of flavor, smell, crunch, mouthfeel, and so on."

One of these machines is devoted to parsing food aromas. It has an appendage that looks like a large plastic bird beak, which Holz-Schlesinger straps over her nose. "You can analyze individually the thousands of different components in the aroma of a tasty burger," she says—components as varied as the scents of "honeydew melon, caramel, cabbage, and dirty socks." Deconstructing the flavors and aromas of beef is just the beginning—Holz-Schlesinger is also working on similar analysis of poultry, fish, and pork.

For all of Impossible's R&D prowess, it's the comparatively quaint products of Beyond Meat, made from extruded peas, beans, and soy, that are currently making their way into the largest number of U.S. households. CEO Ethan Brown (no relation to Patrick Brown, the Impossible CEO) has doubled Beyond Meat's sales annually for the past three consecutive years; his products are now in twenty thousand grocery stores,

Impossible Foods produces faux beef by the hundreds of thousands of pounds per month

including Kroger, Walmart, and Target. Instead of trying to make it taste exactly like meat, Brown says he's trying to make it "just taste good unto itself, and with lots of added benefits." He calls his Beast Burger, which is largely a blend of powdered pea proteins and sunflower oil colored with beet juices, "the ultimate performance burger, with more protein than beef, more omegas than salmon, more calcium than milk, more antioxidants than blueberries, plus muscle-recovery aids." A White Castle burger it is not.

Personally, I'm a fan of the Beast Burger—while it doesn't taste like beef, it sears and grills well and is pleasantly nutty-tasting with great chew. Even so, when I served it on a bun with all the trimmings for my nine-year-old, she saw through the ploy after a few bites. "This is one of your tofu things, Mom. Pass the ketchup."

IN THE AUTUMN of 2018, a few weeks after I interviewed Tyson Food's Tom Hayes, he suddenly and unexpectedly resigned as CEO. The official word was that he did so for "personal reasons," but I had a hard time believing it. The broader meat industry was in turmoil because of oversupply and trade wars with China, Tyson stock was declining, and I had to wonder if the company's shareholders saw Hayes as a man with a premature, or at least a poorly timed, vision. Yet among the dozens of people I interviewed about the future of meat, it was Hayes who made the timeliest and most convincing case for meat alternatives—and cellular meats in particular. He emphasized that the entire "cell-to-fork" process for growing and harvesting lab meats is two to six weeks—a blink of an eye compared with the two and a half years it typically takes to grow cattle from conception to maturity. That represents huge cost and energy savings.

Hayes also pointed out that cultured meats eliminate concerns about *E. coli* and other pathogens that can contaminate animal meat during processing. The single biggest risk in his business, he said, is contamination. A few months after Cargill invested in Memphis Meats, it recalled 130,000 pounds of minced beef that had been contaminated with *E. coli*—a problem that wouldn't happen with lab-grown meat. Uma Valeti describes a test performed in which scientists observed the rate of decay of conventional meats, organic meats, and lab-based meats: left at room temperature, the conventional meats were completely spoiled in less than forty-eight hours; after four days, the lab-grown meats had barely decomposed because there was no trace of bacteria.

Hayes made the case that cultured meats can be produced anywhere—most likely, in facilities near city centers—so they won't have to be transported long distances in refrigerated trucks. For that matter, they'll need less refrigeration because they're less prone to spoilage. Hayes also stressed the potential human health benefits. Lab meats can have the beneficial nutrients of meat—iron, vitamin B_{12}, selenium, niacin, and so forth—while also cutting down on the bad stuff, namely the cholesterol and fat content, by integrating only as much as is needed for flavor. Valeti says that for patients at risk of heart disease, he could conceivably make a beef product with *good* fats—beneficial omega-3 oils, for example—instead of saturated fats.

Hayes shrugged off my concerns that parents wouldn't want to feed lab-grown meats to their kids. His take: it's the safest meat you can produce, and moms and dads will like the fact that their kids won't complain about the flavor of ersatz alternatives. They'll also like the health benefits, the environmental benefits, and eventually, as prices go down and scale goes up, the cost benefits. "If we can make the meat without the animal, why *wouldn't* we do that?" Hayes reasoned.

For all the arguments Hayes made in favor of alternative meat products, he said that he "can't imagine a world where there aren't animals raised and used for human consumption—in my lifetime anyway." Nor, for that matter, can Valeti, who emphasizes the important relationship between animals and the land, and the ecosystemic value of practices like managed grazing. Valeti sees lab meats as an alternative to inhumane, polluting, mass-scale meat production—not as a means to replace high-quality "craft meats" like Sam Kennedy's.

Valeti also recognizes that livestock are culturally and nutritionally important to small farmers, particularly in developing economies. On the farms I've visited in India and eastern Africa, goats, sheep, pigs, and cattle play crucial roles in the agro-ecological system. These livestock convert grasses and agricultural waste into fuel, fertilizer, and high-quality nutrients for populations vulnerable to hunger. They can also serve as a store of wealth, collateral for credit, and a nutritional safety net during times of food scarcity.

For these and other reasons, the argument we tend to hear in Western environmental circles that "meat is bad" is an oversimplification. Both Hayes and Valeti see animal-free proteins becoming "a substantial part" of the $200 billion market for meat—eventually. But in the foreseeable future, they won't be the entirety or even the majority of it. Hayes also asserted that a variety of approaches—plant- and cell-based—to alternative meat production will be crucial to the sector's long-term success. "Just as you see many different electric car models on the market right now," he told me, "there won't be a silver bullet. Customers love choice."

Transitioning to sustainable meat production will take years if not decades, which is why it's as important in the near term to improve animal welfare practices at companies like Tyson as it is to develop alternatives. For my own part, I've resolved to get serious about meat moderation. I plan to rely more heavily on plant-based proteins and save the brisket and BBQ ribs for a blue moon. I'll continue to support Sam Kennedy and other meat producers and I'll be ready to welcome lab-based meats into my kitchen when they're affordable. Sure, I still feel a little squeamish about the live cells twitching in bioreactors, but I'll get over it, just as I've gotten over many of my concerns about GMOs and, for that matter, about sustainably farmed fish and vegetables grown without soil and sun. When I really think about it, cultured meats make me a lot less squeamish than what goes on inside a slaughterhouse or the specter of millions of cloned cattle.

My journey so far—into new and strange frontiers of vegetable, fruit, grain, fish, and meat production—has convinced me that feeding humanity sustainably in the coming decades will require not just major advances in technology, but also the discipline of applying those technologies wisely and equitably. It will require, too, that we look beyond the direct activities of agriculture, beyond the tilling of land and the raising of livestock, to include more abstract and peripheral challenges that are no less urgent, like the development of drought-proof water sources and smart distribution networks that can help irrigate the most parched and populous regions. It will require the preparation of emergency response programs for the famines we can't prevent. It will require, as well, a shift in attitudes and behavior—in particular, around the issue of waste.

Cellular meats represent one approach to waste prevention—they can reduce the resources that go into growing the half of the cow, pig, or chicken that's not eaten. They can also eliminate the waste associated with meat contamination and the rot and cost associated with long-distance meat distribution. The problem of food waste, though, is much bigger than this. Americans waste more food than any other population on the planet. Landfill and agriculture data shows that about 40 percent of the food we produce on U.S. farms rots in fields or fridges or is dumped in the trash. Solving the problem of food waste presents a huge opportunity to feed more people while demanding less of nature.

CHAPTER 9

Stop the Rot

Earth provides enough to satisfy every man's needs,
but not every man's greed.
—GANDHI

IT'S A DRIZZLY March morning in Nashville, and the sky looks like the garbage dump beneath it—a vast gray-brown morass. Against this backdrop, Georgann Parker appears like a Mad Max desperado. She's wearing safety glasses and surgical gloves, tall rubber wellies over her jeans, a bright orange vest over her jacket, and a hard hat over her cropped gray-blond hair. "We can be glad it's not hot," she says, smiling behind her respiration mask. Parker is uncannily upbeat for a woman about to perform a diagnostic exercise that's technically called a waste audit but in Kroger inner circles is referred to as a "Dumpster dive." Parker is Kroger's corporate chief of perishable donations, a role that occasionally involves ripping into hundreds of garbage bags to manually investigate their rotting contents.

Parker has come with two Kroger employees and two officials from Waste Management, the company that collects and dumps all of Kroger's trash. They stand and watch as a compactor truck unloads on the ground in front of them. The trash mound has been generated over the previous six days by a Kroger supermarket located a few miles from my house—one of 117 Krogers in the Tennessee division and one of 2800 Kroger supermarkets nationwide. In all, Kroger's stores serve 9 million individual American shoppers per day and 60 million American families per year, more than a third of the U.S. population. Each store produces many tons of trash a week—most of it perishable fruits, vegetables, meats, dairy, and deli products that have passed their prime or reached their sell-by dates but are still safe to eat.

It's Parker's job to help rescue Kroger's sizable trove of safe-but-unsellable food across its many stores. She oversees the 23 division heads nationwide who manage the company's food rescue operations. Annually Parker and her team capture about 75 million pounds of fresh meats, produce, and baked goods before they get thrown out, and donate them to local food banks and pantries. The number is big, but it's a fraction of Kroger's total fresh-foods waste stream. The company has pledged to donate more than ten times that amount as part of the Zero Hunger, Zero Waste campaign it launched in 2018. The goal is to eliminate food waste from its stores by 2025, and alleviate hunger in the communities surrounding these stores in the same time frame.

"Crazy-big" is how Parker describes the scope of Kroger's goal, "and, yeah, a little daunting. The logistics of food rescue are very complex." For a company as big as Kroger it means not only rescuing the food before it goes bad but coordinating the donations with tens of thousands of food banks and soup kitchens across the country.

Parker, who has round cheeks and pale blue eyes, grew up in a town in Michigan—in a place similar to Garrison Keillor's Lake Wobegon, the fictional land "where it's always pleasant." Parker is that rare person who manages to come off as both effusively and effortlessly friendly. In her colorful midwestern patois, people are "folks," soda is "pop," a lot of something is "a crud-load," and excitement is often expressed as a rhetorical question: "How cool is that?" In other words, she's a gold mine of enthusiasm about things that many of us have trouble caring about—but should.

In the decade before her move to Kroger, Parker clerked for a federal judge in Michigan, and then spent several years serving as a federal

Georgann Parker

probation officer. In this role, she wrote the sentencing recommendations for convicted felons, and she did her best to be positive about even this. "I appreciated the responsibility I had to be sure everybody gets a fair shake under the law, but after a good long run of it, the job beat me down." She sees a continuity between her past and current jobs—"I'm still trying to help fix broken systems," she says. Her law-and-order training has also inspired the moniker she's been given by Kroger colleagues: "food waste sheriff."

Whether because of her chipper disposition or because she's seen more sobering circumstances in her professional past, Parker is unfazed by the disgusting conditions at the garbage dump. Odorous methane emanates up from the depths of landfill below; dozens of overfed vultures circle heavily in the sky above as Parker and her team paw through the mountain of waste. They find bags of potatoes and cabbages that look perfectly fresh, heads of lettuce that look less so, heaps of boxed salads and spinach, serving trays of cut fruit, countless crates of cracked eggs, dozens of meal kits filled with prepped fresh ingredients for cooking shrimp scampi and chicken à la king, packages of sliced salami and cheese, cracked bottles of tomato sauce, dented tubs of icing and ice cream, and cans of Gravy Train dog food marked "Reclaim."

Eventually, Parker becomes sullen. "Gosh, it turns your stomach—

I mean not the garbage, the *waste*," she says. It's the many gallons of milk lying on the ground that upset her most. She checks their sell-by date—eight days left before they expire. "There's no good reason for this." Next to the milk is the carrion the vultures are after: packages of breakfast meats and pork chops, steaks, minced beef, a ham hock, and a heap of twenty or so rotisserie chickens, roasted the color of the mud in which they rest.

Parker and her team separate the waste into categories, weighing and photographing the contents of each. Later, they'll crunch the numbers and find that more than half—52 percent—of the waste produced by the supermarket could have been donated, recycled, or composted. They will assemble the data into graphs and charts with crime-scene-esque photos. "You can see which departments in the store are doing their job of food rescue, and which are not," Parker grumbles when we discuss the results. "A lot of this should have been given a second chance."

FIFTY-TWO MILLION TONS of food are sent to U.S. garbage dumps annually, and another 10 million are discarded or left to rot on farms, according to Darby Hoover, a waste researcher with the San Francisco office of the environmental group Natural Resources Defense Council (NRDC). Put another way, Americans waste enough food to fill a 90,000-seat stadium every day, and that's about 25 percent more per capita than we were wasting in the 1970s. Most of the food waste in the United States, about 35 percent, is generated by households. The average American throws out more than a pound of food a day—some 400 pounds per year each. Restaurants and retailers like Kroger are close behind, generating another third of it. The value of the food wasted in America each year has been estimated at between $162 billion and $218 billion.

Hoover sees the problem from an environmental angle. "Wasting food also means wasting all the water, energy, agricultural chemicals, labor, and other resources we put into growing, processing, packaging, distributing, washing, and refrigerating it," she observes. The nonprofit group ReFed estimates that food waste consumes 21 percent of all freshwater, 19 percent of fertilizer, 18 percent of cropland, and 21 percent of landfill volume in the United States. Add to that the methane problem: only 5 percent of food waste in America gets composted into soil fertilizer, using a controlled process in which bacteria and heat decompose food scraps into rich plant nutrients. The other 95 percent of food waste goes to landfill and rots in an uncontrolled way, emitting methane, a potent greenhouse gas. "If food waste around the world was a country, it would rank third behind China and the U.S. in terms of greenhouse gas emissions," says Hoover.

To Georgann Parker, the problem is a social injustice: "Wasting food—especially healthy perishables—becomes an ethical problem when you consider that there's about forty million folks in this country living in poverty who don't have reliable access to nutritious food." Less than a third of the food we're tossing would be enough to feed this underserved population.

Kroger has half a million employees, many of whom are paid minimum wage and themselves face food insecurity. Parker says, "Having a larger purpose that sustains low-income communities—that builds employee morale and it honors the company's traditions." Kroger's founder, Barney Kroger, started with a bakery business and he used to hand out day-old bread and pastries every evening to low-income neighbors. "Our brand and core values hinge on food accessibility," Parker maintains.

Kroger's Zero Hunger, Zero Waste campaign is also a bottom-line opportunity for the company, which has to pay increasingly steep "tipping fees"—the costs imposed by some state governments on the loads of waste that a company or institution dumps. Kroger can also collect millions of dollars in federal tax breaks annually for its food donations. Pressure to cut waste is also coming from Kroger's investors. Nearly every major brand in food retail, including Publix, Walmart, Costco, Target, and Whole Foods, has introduced waste-reduction programs in the past five years. In a recent assessment of these programs by the

Center for Biological Diversity, a nonprofit based in Tucson, Arizona, Kroger was the third-highest performer in the not-very-high-performing bunch—it scored a C on the overall grading scale.

Food retailers have historically been lax on waste management, but they're under increasing pressure to reform. When Amazon, for example, resisted efforts to address the food-waste problem within Whole Foods and its other retailers, top investors said they'd pull out if the company didn't start reining in waste. Kroger CEO Rodney McMullen got similar prodding from BlackRock, his biggest investor. Jessica Adelman, Kroger's head of corporate affairs, says, "You gotta believe when our largest shareholder tells my boss: If you don't have a social-impact proposition then we're not going to continue to invest in you—that resonates."

Adelman enlisted World Wildlife Fund (WWF), which conducts a large food-waste research program (on the grounds that agriculture is the world's biggest threat to wildlife habitat), to help Kroger devise a strategy for food-waste prevention and donation. WWF encouraged the Dumpster dives and other rigorous waste-stream analysis. "There's a misconception that the answer to food waste is composting," says Pete Pearson, WWF's director of food-waste research. "The real emphasis—whether you're a company, or a household, or a city—needs to be on prevention first, then rescue and donation, then composting as a last resort."

What makes solving food waste so difficult, says Pearson, is that "there's no single technology or policy intervention that can nip this thing in the bud." He continues, "The problem occurs upstream and downstream, in fields, warehouses, packaging, distribution, supermarkets, restaurants, and homes." It will require participation not just from a few "food waste sheriffs" like Georgann Parker in the corporate trenches, but from an army of them at many levels of the private and public sectors: academics and federal policy makers who are trying to standardize expiration dates and incentivize food rescue; players in city and state government building curbside composting programs and increasing "tipping fees" to discourage waste; software developers building apps that connect people with food surpluses to people with food deficits; materials scientists finding new ways to preserve perishable foods and extend shelf life; engineers designing machines to accelerate large-scale composting; and activists leading campaigns to transform public consciousness on this issue.

Over several weeks, I ventured behind the scenes of Kroger to try to understand the nuts and bolts of a zero-waste strategy, following that process through the three stages—prevention, rescue and donation, and ultimately composting. First I needed context—a better understanding of why we squander so much food in the first place.

"THERE ISN'T ANY waste in nature. Anything that dies in nature becomes food for something else," Darby Hoover tells me. "Humans have created waste as a concept, and we should be able to *uncreate* waste as a concept."

Hoover recently conducted a two-year study exploring and comparing food-waste patterns in three U.S. cities—Denver, New York City, and, coincidentally, my hometown, Nashville. "There's surprisingly little hard data about who's wasting what, where, and why, and that makes it harder for cities and companies and households to solve this problem," she asserts.

To run her analysis, Hoover worked with Tetra Tech, a San Francisco–based engineering company that specializes in waste logistics. They recruited 1150 residents in the three cities to participate, all of whom agreed to offer up their trash for inspection. More than half kept "kitchen diaries," noting when they threw out their food and why. The foods most often dumped in the trash or poured down the drain included brewed coffee and coffee grounds, bananas, chicken, apples, bread, oranges, potatoes, and milk. Hoover noted the conspicuous absence of things like Doritos, Spam, and Twinkies on this list. "Food waste is riddled with unexpected contradictions, and one of them is that healthier diets tend to be the most wasteful diets," she says. "Our current cultural obsession with eating fresh foods is a great thing from a health perspective, but not so great from a waste perspective."

Hoover also found that in Denver and New York, which both have municipal composting programs, the participants who regularly composted their leftovers tossed out significantly more food than noncomposters, presumably because they felt better about the outcome of the waste. "Prevention is the holy grail of waste work," says Hoover. "*Not* generating waste is far better for the planet than recycling food scraps." She also found that parents with young kids generated waste in their efforts, however hopeful and virtuous, to expose their kids to new flavors and healthy offerings—only to have the kids refuse to eat it. The upshot: food waste at the consumer level "is often tangled up with good intentions—and that makes it particularly tricky to solve," Hoover observes.

I'll admit that too much of the food in my own household goes uneaten. My husband distrusts anything that's been in the fridge for more than two days. I like to buy those big bundles of fresh spinach and large bunches of bananas in hopes of plying my family with nutrient-dense foods, only to have the bananas brown and the spinach molder. I chronically overcook for guests and get excited about new recipes for which I buy too many esoteric ingredients—only to use a teaspoon of this and a quarter cup of that. Weeks later, when no one's looking, I guiltily throw the neglected remainders away.

Beyond good intentions, several other factors pave the road to the local dump here in the United States. For one thing, there's what Hoover calls "our conformist standards of beauty." American shoppers have a very rigid idea of what fruits and vegetables should look like, she says, and it "doesn't include produce that's marred or that's grown in irregular shapes, or that's gotten bruised, browned, bumped, wilted, or discolored during their journeys from field to market." The average American shopper has aesthetic standards not unlike Tony Zhang, chucker of gnarled carrots: we reflexively snub irregularities. The problem lies both with the shoppers who demand perfect-looking foods, and with the grocers who behind the scenes reject the nonconformist fruits and vegetables. "Unthinkable quantities of fresh produce grown in the U.S. never make it to the store because they don't make the aesthetic cut," Hoover asserts. A recent study in Minnesota found that about 20 percent of all the fruit and vegetables produced in the state get trashed because they don't meet our narrow aesthetic standards.

A heap of rejected bell peppers in Salinas, California

Common victims: table grapes that don't grow as a wedge-shaped bunch are left to rot in the field or hauled directly from field to landfill. The same goes for lopsided bell peppers, gnarled carrots, blemished apples, and so forth. (Think back to Andy Ferguson's frost-bitten apples—perfectly healthy and delicious, but the ones with frost rings won't sell.) Large organic vegetable farmers, said Hoover, routinely toss more than conventional growers because their products are less uniform. The unfortunate irony is that marred produce is often more nutritious and flavorful than unmarred produce—fruits and vegetables in fact produce flavors and antioxidants when they're under stress from insects, heat, frost, or blight.

Hoover, who built Stanford University's first recycling program as a student there in the early 1980s, wrote her thesis in graduate school on the psychology of waste. The desire for perfect produce has been around since long before the emperor Tiberius was demanding perfect, year-round snake melons, but Hoover maintains it reached a new level of intensity in the United States in the 1950s, as housewives adapted to widespread refrigeration, new packaged products, and internationally shipped fruits and vegetables. "Suddenly, you could eat pineapples in

Maine and strawberries in January. It was the era of Wonder Bread, and TV dinners," she says. "Perfect, rote foods came to represent safety and innovation." Today, this obsession has reached still further heights owing in part to the "camera cuisine" trends on social media. Hoover and I discuss the Instagramming of golden fresh-baked pies and arty restaurant entrées—a feel-good phenomenon that, to her mind, reinforces a cultural obsession with perfect food and a tendency to reject anything less.

EVEN AS AMERICANS blithely ogle #foodporn, Europeans are learning to think in more realistic terms about food value. Selina Juul isn't a politician, yet according to the Danish government she's largely responsible for reducing that country's food waste by 25 percent in five years. Juul was born in Russia and moved to Denmark in 1995, when she was thirteen. "I come from a country where there were food shortages. We had the collapse of infrastructure, communism collapsed, we were not sure we could get food on the table," Juul told the BBC. "Then I was really shocked to see a lot of food getting wasted."

She had an interest in graphic design and began waging clever public campaigns. Juul founded Stop Spild Af Mad (Stop Wasting Food) as a Facebook group in 2008; it now has tens of thousands of followers and has shown up in supermarket boardrooms, on the TED stage, and in the European Union Parliament advocating for waste reform. "Food waste is the lack of respect for our nature, for our society, for the people who produce the food, for the animals, and a lack of respect for your time and your money," says Juul. She helped rebrand restaurant doggy bags as "goodie bags," and distributed sixty thousand of them across the country. Denmark's supermarkets started selling bananas with single-item discounts under a sign that read TAKE ME I'M SINGLE, reducing banana waste by 90 percent.

The trend caught on: Wefood, a Danish charity, opened what it called "the world's first food-waste supermarket" in Copenhagen, selling rejected produce and food nearing its best-by date. Nine months later it opened a second branch. The country's major supermarket chains stopped offering quantity discounts that entice shoppers to overbuy; many have added "stop food waste" sections, where they aggregate the older, cut-rate offerings.

The Danish momentum rippled elsewhere. In London, activist Adam Smith founded the Real Junk Food Project and opened the country's first food-waste supermarket along with a chain of "pay as you feel" cafés that cook up soups and sandwiches from ingredients that would've been tossed at no fixed price. Similar cafés have sprung up in Australia and Israel.

Another promising London-based effort is "Olio," a "food-sharing app" that connects not just businesses and food banks but also individual neighbors to one another to off-load their excess food. "Ever cooked too much for dinner? Ever bought a pack of onions and only need 1? Going on a holiday and your fridge is full of food?" reads the Olio website. The endeavor was slow-going when the app debuted in 2016, but by 2019 it had over half a million members—mostly neighbors sharing the contents of their fridges. A Copenhagen-based app, Too Good to Go, has also had success selling users discounted just-before-closing bakery and restaurant food. The British supermarket chain Tesco has moved to recyclable packaging on 85 percent of its own-brand inventory and eliminated best-by dates on its products altogether, encouraging customers to trust their own judgment.

Not to be outdone, France recently passed a law banning grocery stores from throwing away unsold food, threatening fines of up to €3750 for an infraction. In some French cities, a fleet of "food ambulances" collects waste from grocers and stores and delivers it to churches and synagogues. The shift in public consciousness has pushed the European Union to establish a goal of cutting per capita food waste in half among retailers and consumers by 2030.

We may be a long way from seeing this kind of shift occur in the United States. For one thing, everything is bigger here—our shopping carts, plates, portions, appetites, and of course our people. "If you're willing to indulge a little food-waste psychoanalysis," says Hoover. "Americans irrationally associate the size and quantity of food we consume—and the waste we generate—with the extent of our freedom and power." We get away with this in part because food is relatively cheap in America (aided by subsidies to crops like corn and soybeans). Middle- and upper-income families in the United States spend a much smaller portion of their household budgets on food than nearly anywhere else in the world. When I ask Hoover for some advice that could help us prevent household food waste, she gives me marching orders.

First, leftovers should be fine to eat for at least a week (she stretches hers to ten or more days and has never gotten sick). "Use your eyes and nose," she exhorts. "If it looks and smells fine, eat it." If you can, use glass storage containers, which keep food fresher for longer than plastic. Buy mottled or misshapen produce: it tastes just as good and it's probably better for you than the perfect-looking stuff. Choose frozen fruits and veggies over their fresh counterparts—they won't go bad and they're no less nutritious (some nutrients do get lost when foods are blanched before freezing, but others are preserved because the foods get frozen right after harvest and don't deteriorate in transit to market).

Activist Tristram Stuart of the London-based group Feedback in Trafalgar Square at a feast made from reclaimed food for five thousand people

Lastly, "channel your grandmother and reimagine your leftovers," says Hoover: Sunday's roast chicken can become Monday's chicken tacos and Tuesday's tortilla soup.

Pete Pearson of WWF, for his part, supports both traditional and tech-forward approaches to food waste prevention. He cites the example of Opal apples, a non-GMO version of the contentious Arctic apples that are gene edited to eliminate browning of the flesh. There are also CRISPR-edited nonbrowning mushrooms coming onto the market, and potatoes that have been edited to be less prone to browning, bruising, and black spots—meaning fewer will end up in landfills.

"HERE'S WHERE WE put the uglies," says Georgann Parker as she walks me through the produce section of a Kroger supermarket outside of Indianapolis, Indiana—one of the largest stores in the chain. I've come to get a crash course in supermarket logistics and a glimpse into the company's waste-prevention efforts. I'd never noticed these particular offerings in my local Kroger before. There, tucked into the side of an island bearing those iconic pyramids of supermarket fruit—perfect orbs of red, orange, green, and yellow—is a four-tiered shelf topped with a sign MARKDOWN! BEAUTY IS ONLY SKIN DEEP. The shelves bear mostly empty straw baskets of gnarled bell peppers, arthritic-looking carrots, too-small cantaloupes, and cucumbers curved like pistols.

While fresh produce accounts for less than 15 percent of Kroger's profits, it's in the company's interest to sell every last misshapen product. "We want everything that comes in the back door to go out the front, but of course it doesn't," says Parker. Most of the ugly produce is rejected before it gets to a store and left to farms to dispose of, but inevitably some misshapen fruits and vegetables make their way into the shipments. "They used to get immediately tagged for donation to food banks, but now we're selling them at a steep discount, and they almost always sell out," she tells me. There's some waste they simply can't sell—because stores overorder, or because refrigerators fail, or because customer purchasing patterns shift and aren't what buyers predict.

Kroger introduced the "uglies" section into its stores in early 2017, around the time that activists and entrepreneurs were embracing cast-off produce. A start-up called Imperfect Produce launched a subscription delivery service for "funky fruits and vegetables" in the San Francisco Bay Area and began selling its irregular products in Whole Foods. Hungry Harvest, Ugly Mugs, and Food Cowboy are among other new enterprises that have been building markets for millions of tons of rejected produce. "These efforts are seeing gradual success, but they're rerouting only a small fraction of the rejected produce that's still safe to eat," says Pete Pearson. "It'll take the big players to really move the needle."

Pearson is working with Kroger on a plan to capture the aberrant produce earlier in the waste stream, not for direct sales but rather for the enormous amount of prepared and packaged foods—potato and macaroni salads, coleslaws, pizzas, frozen fruits and vegetables, and so forth—that Kroger makes and sells through its eponymous and private-label brands.

The ugly produce program works in tandem with Kroger's regular markdown programs. If meats don't sell within a day of their sell-by dates, they get pulled from shelves, slapped with a "WooHoo! *MARKDOWN*" sticker, and placed in the sale area of the meat section. If the discounted products still don't sell, they're supposed to be yanked the night before their sell-by date, scanned out of the system as a loss, and put in a backroom freezer for donation. A similar process is supposed to be followed for bakery items and dairy products. The company policy is to pull milk from the dairy case ten days before its expiration date, at which point it can be donated fresh, or frozen and then thawed for donation. "There's no good reason any milk products sold in a Kroger store should ever be dumped," says Parker.

Confusing sell-by labeling is another major barrier to waste prevention both in supermarkets and in homes. The dates printed on the perishable products you buy are not federally regulated and do not represent any technical or standardized measure of food safety. The Food and Drug Administration, which has the power to regulate date labels, has chosen not to do so because no food-safety outbreak in the United States has ever been traced to a food being consumed past date. (They've been traced instead to certain pathogens that may have contaminated the food during processing; or to "temperature abuse," like leaving raw chicken in a hot car; or to air exposure that encourages mold.) "You're far more likely to get sick from something because it's contaminated or gone unrefrigerated than because it's past-date," says Parker.

Milk has the most inconsistent labeling, state to state. Most milk is pasteurized, a process that eliminates the risk of food-borne illness, even after the sell-by or use-by date. The sell-by date that dairies generally print is twenty-one to twenty-four days after pasteurization. However, Parker tells me, "milk that's been properly refrigerated is safe to drink well after that." Some states like Montana impose even stricter time limitations, requiring sell-by dates of just twelve days after pasteurization, and ban the sale or donation of milk after that date, which wastes countless gallons of good milk.

"Supermarkets have to juggle dozens of different date-labeling laws and they lose about $1 billion a year from food that expires in theory—but not in reality—before it's sold," says Emily Broad Leib, director of the Food Policy Program at Harvard Law School. "Date label confusion harms consumers and food companies, and it wastes massive amounts of food." Leib helped develop the Food Date Labeling Act, proposed federal legislation that would standardize labels to "best if used by," a phrase indicating that a product may not be optimally fresh but is still safe to eat. The bill will also prohibit states from preventing stores or manufacturers from donating products that have passed their peak-quality periods but remain edible and nutritious.

The date-labeling effort is part of a broader Food Recovery Act legislation, which was introduced in the Senate in 2017 and would standardize date-labeling and food-rescue laws throughout the fifty states.

It would also incentivize state schools and government institutions to make use of ugly fruits and veggies that never make it to market.

Georgann Parker says Kroger is throwing its lobbying weight behind these laws while also pushing for better food packaging. Materials scientists are now finally beginning to break new ground in food packaging and preservation techniques. The challenge in preserving freshness of perishable foods comes down to sealing out oxygen. It's a seemingly benign gas, but when it penetrates food packaging it feeds mold growth and speeds the proliferation of microorganisms and enzymes. Especially in foods untreated with chemical preservatives, oxidation can degrade the flavors, pigments, and textures, and the quality of nutrients, oils, and fats.

Kroger has partnered with Apeel Sciences, a Silicon Valley start-up founded in 2012 by a young materials scientist, James Rogers. Rogers studied the casings—skin, rinds, and peels—that fruits and vegetables naturally create to seal out oxygen and protect themselves from decay. The big idea, he says, is "using food to protect food." He found a way to recycle organic ingredients, like grape skins left over after wine pressings, to create a natural sealant that can be sprayed on fruits and vegetables and extend the shelf life of produce up to three times longer than conventional produce. The film is transparent, flavorless, and completely natural, and debuted in 2018 as a protective casing on avocados in the produce sections of Krogers throughout the Midwest. It's a good example of third way thinking: "We don't need to go into a lab and create new chemistries to solve old problems," says Rogers. "We can draw inspiration from plants."

Chemists in labs are also making important progress. Parker tells me that researchers are developing oxygen-absorbing films that can be incorporated into either flexible or rigid packaging materials, and reduce the oxygen concentration within to less than 0.01 percent—more than doubling shelf life. The problem is cost. Food manufacturers are still packaging bread in the same plastic bags and eggs in the same cardboard cartons they've been using for aeons because it's cheap.

Kroger's investment arm is funding start-ups that are developing new packaging technologies, but Parker maintains we need a coordinated industry-wide R&D effort on this front.

Pearson of WWF is optimistic about advances in data-management tools that consumers can't see. Using product ID codes and tracking systems like Blockchain, Kroger can monitor the movement of each one of the billions of products that flow through its stores, as well as the shopping habits of 60 million families. All this data is managed with the goal of synchronizing each store's supply with its shoppers' demand—thereby shrinking the volume of unwanted and expired products.

There will always be some amount of oversupply and therefore waste in supermarkets, says Pearson, "until the day we create a Star Trek replicator"—which would enable food to materialize upon demand. In the meantime, digital tools will get ever-better at tracking the life cycle of a product from conception to sale, which will go a long way to helping supermarkets reduce their excess inventory and donate far more of it.

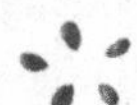

I DROVE WITH Georgann Parker twenty minutes east of Indianapolis to see the final stage of Kroger's zero-waste strategy at K.B. Specialty Foods in Greensburg, Indiana. This is one of thirty-seven food-manufacturing plants owned and operated by Kroger. It produces 90 million pounds of food a year in giant stainless-steel vats, mostly deli products like potato and pasta salads, coleslaws, relishes, mac and cheese, and so forth that are sold in Kroger's stores.

What sets the K.B. Specialty Foods facility apart from other Kroger plants, aside from the strange hodgepodge of products it manufactures, is that the raw ingredients in those products, which include 200 million russet potatoes, 16 million cabbages, and 7 million pounds of powdered cheddar cheese per year, create spectacularly bad-smelling waste. "All those potato peels and cabbage cores stink pretty bad if they're sitting in garbage bins on a hot summer day," says Parker. The manufacturing facility is located in a residential area, right next to an elementary school, and in 2016 the neighbors began to complain about the stench.

The solution proposed by the plant-operations director was an anaerobic digester, which is essentially an industrial-scale composting system contained inside a sealed tank. From the outside, the digester at the K.B. facility looks like nothing but a circular forty-foot-tall water tank. Inside, it works like a biochemical stomach that uses enzymes and microorganisms to break down the organic matter—food scraps combined with wastewater—in an anaerobic, or oxygen-free, environment. The microorganisms can break down not just fruits, veggies, and starches, but also meats, fats, oils, and grease. Instead of producing a soil fertilizer, as outdoor composting does, the digester system converts food waste into biogas—a fuel that generates heat and electrical energy for the plant.

Anaerobic digestion is hardly a new concept—it's occurred naturally for millions of years. The first anaerobic digester was built in India in the nineteenth century, but only in the last decade have engineers in the United States begun to update these systems for use at industrial sites. There just hasn't been demand for such systems—it's far cheaper for K.B. Specialty or any other food producer to dump its waste in landfills. But now as landfills fill up, concerns about methane emissions escalate, and the cost of food waste increases, this option is beginning to make more sense.

The Seattle-based start-up WISErg has raised $70 million to produce its small-scale anaerobic digester designed to be used at restaurants, grocery stores, schools, hospitals, and other community centers. An average-sized unit can process 4000 pounds of food scraps daily and eventually pays for itself given its sellable end-product: natural biogas, a clean, local renewable energy resource. The Massachusetts-based start-up Harvest Power has also raised tens of millions of dollars to build composting technology, which it markets with the motto "VISUALIZE WHIRLED PEAS."

Such an investment may make economic sense in a place like San

Food scraps become soil at a composting site in Nashville, Tennessee

Francisco, the first city in America to ban food waste. Since 2007, every household, business, and public facility in the city has been required by law to participate in Recology, San Francisco's composting and recycling service. Those that don't are fined. In the last decade, hundreds of cities, from Bridgeport, Connecticut, to Boise, Idaho, have introduced voluntary municipal composting programs, many with curbside pickup of food scraps. Nearly a dozen of these have, like San Francisco, outlawed food waste altogether, imposing penalties on companies and households that don't participate.

Outdoor composting is a natural form of aerobic (meaning oxygen-aided) digestion. Trillions of tiny microbes inside a compost pile, fueled by oxygen, eat the food and yard waste, excreting nitrogen-dense fertilizer that enriches soil. In San Francisco, among other cities, municipal compost is redistributed to farmers in the region and spread across their land to both nourish and protect it, sealing in moisture and shielding the soil from drought.

Composting programs and anaerobic digestion systems complement rather than compete with each other—the former works on a large scale in and around cities, while the latter works well at specific sites in urban and residential areas where you need maximum odor control. The Environmental Protection Agency estimates that, all told, Americans are currently producing about 23 million tons of food and yard waste compost a year. "Eventually we'll get to a point where every city and township in the United States has a mandatory composting program, and every supermarket, restaurant, and food plant converts its waste to energy or animal feed," says Pete Pearson. "The hope is that our grandchildren will consider food waste as archaic as snail mail and corded phones."

There's little question that implementing zero-waste strategies in cities, businesses, homes, and public organizations will be crucial to third way food production. Nashville, to take one example, has set a goal of becoming a "Zero Waste City" by 2030—one of hundreds of U.S. cities that have made a similar pledge. These goals aren't binding, and Nashville's plan is undoubtedly vague, but Pearson sees it as a grassroots shift in the right direction.

Solving food waste will transform a linear food system into a circular one, says NRDC's Darby Hoover: "Linear economies are based on consuming, depleting, and throwing out. A circular economy is designed on growing, reusing, and regenerating resources." This concept of circularity is as old as time, and it's still present in subsistence farming systems the world over, but it got engineered out of the industrial food system, Hoover says, and "now's our chance to engineer it back in."

Doing so will require more than composting programs, smart apps, and ambitious federal policy. We'll need to see a shift in public consciousness and a willingness to participate on an individual level. Individuals—you and me—will also need to play a role in solving another fundamental threat to food security in the future: freshwater.

Agriculture consumes 70 percent of the world's freshwater, a resource we waste as recklessly and as unwittingly as the food it's used to grow. The Gandhian principle of need versus greed applies here, too. If we hope to build safe and reliable food systems going forward, we'll need to collectively participate in the creation of what may be the single most valuable resource of the coming century: a drought-proof water supply.

Chapter 10

Pipe Dreams

He looks into his Dixie cup and looks back up as if surprised at what he found there. The future, maybe.

—Wallace Stegner, *Crossing to Safety*

AMIR PELEG HUNCHES his burly six-foot-three frame into a cement tunnel leading to one of several reservoirs that supply water to Jerusalem. Condensation collects and drips from the ceiling, inches overhead, like thousands of tiny stalactites. Peleg catches a droplet on his palm. "*Haval al kol tipa,*" he says, a Hebrew phrase meaning "every drop counts." Located at the edge of the city, this water reservoir is concealed in a massive underground vault patrolled by armed guards to keep adversaries from poisoning the supply. Thick walls of chiseled rock surround a pool of water lit by floodlights, ghostly and luminous, forty feet deep and wider and longer than two football fields. "This is the modern-day Gihon," Peleg says.

Gihon was the ancient spring that made human settlement possible in Jerusalem circa 700 BC. Today, freshwater sources in Israel and the surrounding region are more precious than they were in the Iron Age. About one million residents continually draw water from this reservoir, which is filled by pipelines snaking ninety miles south from the Sea of

Galilee, one of the few freshwater sources in the region. Like most of its neighbors, Israel is a desert nation, but the past decade has brought less rainfall than any other period in at least nine hundred years.

The Sea of Galilee and Israel's other natural freshwater sources are now so overdrawn that they can provide only 10 percent of the country's total water needs. Yet through feats of both thrift and ingenuity, Israel has managed to achieve a freshwater surplus and has found a way to produce higher agricultural yields than it did in nondrought years. The country of eight million is 95 percent agriculturally self-sufficient (importing coffee and other specialty foods, but no staples), and a major exporter of dates, avocados, olive oil, pomegranates, citrus, and almonds.

Farming is a thirsty business, especially if you produce fruits and nuts. It takes a gallon of water to produce a single almond; three gallons per olive; five gallons for a pomegranate; seven gallons per grapefruit; nine gallons an avocado. "Water in agriculture is like blood in the body, or vibration in sound, or the wizard in Oz," Peleg tells me. "It's the essence of the thing. No water—no food."

Agriculture consumes about three-quarters of the world's water supply, and in an intensively farmed country, it's an even higher share—about 80 percent of Israel's water supply goes to its food production. As a nation in chronic political conflict with its neighbors, Israel has spent decades trying to become entirely self-sufficient in its food and water supply. "I don't think it's overkill to say that Israeli entrepreneurs are reinventing how the world makes and manages freshwater," Peleg tells me on our tour of the Jerusalem reservoirs. Peleg has become one of the leaders of a water-tech movement that began in the 1950s and has since produced an array of innovations designed to both use less freshwater and create more of it.

"We can't depend on our neighbors for food and water," he observes. And Israel's neighbors can't depend on them. Israel pumps a portion of its water surplus (about 21 billion gallons per year) to Jordan and the Palestinian Authority, but residents of the West Bank receive less than half the freshwater supply, per capita, than Israelis and have extremely limited arable farmland. About a quarter of Palestinians are food-insecure, and nearly half of the population in the Gaza Strip relies on food aid.

The ethics of water and land distribution in the region are controversial, but Israel's water technology is viewed by its allies and enemies alike as a modern marvel. There are plenty of water-strapped countries in the world, but few if any are so parched and so agriculturally productive. The range of Israeli technologies in development is vast—from microscopic sewage scrubbers to hyperefficient irrigation systems and supersized desalination plants. The foundation for all of these is the so-called smart water network—a system of sensor-embedded pipes that distribute water throughout the country. Peleg is among the entrepreneurs designing the digital nuts and bolts of this system. His self-given job title: "chief plumbing officer."

Amir Peleg

Peleg's company, TaKaDu, designs software that uses mathematical algorithms to spot and prevent leaks and bursts in water pipelines. Detecting leaky pipes may seem like a small concern, but it matters a lot in environments where water is scarce and expensive. "Water is like champagne to Israelis," says Peleg. "You wouldn't pour Veuve Clicquot into a broken flute!"

PELEG IS FIFTY-TWO, with silver buzz-cut hair, arching black eyebrows, and a jaw like an anvil. He is George Clooney's brash Danny Ocean crossed with the affable Schneider from *One Day at a Time*—part swaggering CEO and part scrappy superintendent.

Peleg lives with his wife and their three young kids in an agricultural village about thirty miles outside of Tel Aviv, where during his off hours he tends eight acres of farmland he calls his Eden. He grows olive, pomegranate, avocado, lemon, fig, mango, and pecan trees, a vegetable and herb garden, and a small vineyard with grapes for merlot and chardonnay. Backyard farming is "my most extravagant hobby," he says. Peleg pays thousands of dollars a year to irrigate the farm. On weekends he brines his olives, pickles his cucumbers, and ferments his wine.

Peleg had little exposure to agriculture growing up. His grandfather built Tel Aviv's first luxury hotel and his dad ran it. At thirteen, Amir hacked the first Apple computers that came to the market in Tel Aviv, creating a version with Hebrew characters that he sold to local businesses. At seventeen, he was accepted to Talpiot, the Israel Defense Forces' most elite technology unit. Over eight years, he learned to develop military drone operating systems and software to identify key visual information, such as tanks and missiles, in satellite images.

Peleg then applied his programming skills to large-scale textile production. He built software that analyzes visual data to identify flaws in fabrics. He founded YaData, another algorithmic venture; its software helped online advertisers more accurately target their customers. Two years later, Microsoft bought it for $30 million. TaKaDu is a logical extension of his earlier enterprises. "It all boils down to finding new ways to understand aberrations in data," he says.

Peleg got the idea for TaKaDu at a technology conference in Vienna in September 2008, during a chat with a water engineer. The conversation "shook me to the core," he recalls. Peleg learned that, on average, utilities worldwide lose about a third of the water they distribute through leaks and bursts in their pipelines. In the oldest and most brittle water networks, London's, for example, leakage is about 60 percent. The engineer "told me more than half of the water pumped through the London system gets wasted," recalls Peleg. "This can't be fixed by hacking up streets to find where the problems are." Water networks in the United States, especially in the rural regions that feed American farms, waste up to about 30 percent of the water they distribute.

"I'm standing there and it hits me all at once—the magnitude of this waste. Can you imagine managing a factory where a third of your inventory is lost before it gets to your customer? And that being acceptable?" Peleg balks. "We're doing this with water—at a time when so much of the world is drying up."

The engineer Peleg met specialized in building so-called SCADA systems in Israel's pipelines that collect data from the embedded smart sensors, also known as telemetry. These sensors use mechanical devices, like rotating wheels, as well as ultrasonics, to measure network flow, pressure, and quality, and they can transmit hundreds of data points every fifteen minutes. Peleg wasn't interested in the hardware, just the information it generates. "I asked the SCADA guy what he does with the data. He says, 'We store it.' I thought, This is it! I'm going to mine this data for golden nuggets."

Within a few months, Peleg was running TaKaDu out of his living room. A number of Peleg's early recruits were from the Talpiot program. "I said, 'Now our enemies are not people, but the leaky pipes underground.'"

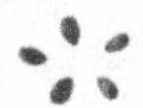

TAKADU'S HEADQUARTERS ARE above a Pizza Hut and a pastry shop, in a glass-and-granite mixed-use office building in a quiet suburb of Tel Aviv. The office looks like a dressed-up college dorm, with minimalist couches and mismatched chairs, walls painted primary colors, and an open kitchen with a large picnic table for meetings and meals. There's a Brady Bunch–style collage of childhood photos of the company's fifty employees and poster-sized photographs depicting water in some form, from dewy fields to foaming falls.

Before showing me around, Peleg invites me into his modest office for a history lesson. Israel's leadership in water technology, he says, dates back to the early 1950s, when Prime Minister David Ben-Gurion made an address to the citizens of the newly established state of Israel now immortalized by four words: "make the desert bloom." The country, just partitioned from Palestine in 1948, was more than 70 percent desertified. Ben-Gurion wanted Israel to lead the regional race to establish self-sufficiency in both its water and food supply.

To spearhead this effort, Ben-Gurion enlisted Simcha Blass, an engineer who produced Israel's first major breakthrough in water conservation technology. Peleg recounts the story of this discovery: "It was in the 1930s that Blass visited a friend near Haifa, and as the two ate lunch outside, he noticed something odd," says Peleg. "In the field before him, he saw what looked like a fence of trees, all of them scrawny except one that was big and lush. The friend said it was a great mystery because there was no river or aquifer nearby. The tree had grown without water." Blass did some investigating and found a tap with a small dripping leak that had over time soaked the root system underneath the big tree.

In the late 1950s, a time when fields were watered by means of intentional flooding and sprinkler systems—practices that still dominate American agriculture today—Blass started playing with different prototypes for irrigation systems made from plastic hoses pricked with tiny holes from which the water could drip. He placed spiral "microtubing" inside to slow the velocity of the water. In August 1965, he tested the technology at Kibbutz Hatzerim, one of the hundreds of agricultural communes that produce Israel's food. The drip irrigation was more than twice as efficient as conventional methods and produced larger crop yields. In January 1966, the kibbutz began manufacturing drippers under the name Netafim, which means "drops of water." Now the company has 4400 employees and sells its wares worldwide, pulling in about a billion dollars a year in revenue.

Two years before Netafim was founded, in 1964, Alexander Zarchin, another Israeli engineer, invented a commercial process for removing salt from seawater. First he froze seawater in a vacuum to form pure, salt-free crystals, then melted them to produce potable water. A year later, Zarchin founded the company IDE with the goal of "transforming the world's oceans into affordable clean water." The company has since developed other methods for desalination and become the world's largest manufacturer of desalination plants. These facilities are able to perform what Peleg half jokingly describes as "an act of God," recalling a passage in the book of Genesis where God makes "the salt water sweet."

It was also in the 1960s that Israeli engineers began to develop early processes for filtering human sewage into recycled wastewater. Israel now recycles more than 85 percent of the water supply that gets flushed down its toilets, sinks, and gutters. The sewage is cleaned using a series of filtration processes that include "bioscrubbers"—bacteria that break down the waste. The result is water that's perfectly safe for growing crops, just not good enough to drink. This reclaimed water is pumped through a vast network of pipes that are painted a bright purple. This "purple pipe network" now feeds water to Israel's farms and factories. A separate network delivers the premium-grade water drawn from freshwater sources like the Sea of Galilee and from high-cost desalination plants. This is the precious "champagne," as Peleg calls it, that runs to domestic taps for drinking, cooking, and bathing.

The desalination and sewage-scrubbing technology that Israel's been using for decades is now spreading into many regions of the world, including the United States. But manufacturing freshwater is expensive—and it makes economic sense only if delivered by highly efficient water networks. "The economics of clean water depend entirely on conservation technologies," says Avshalom Felber, the current CEO of IDE, "and among all of those in development, the most valuable detects leakage in networks."

TAKADU HAS SAVED its customers billions of gallons of water since it was founded in 2008, yet the means by which the water has been saved are mostly hidden from view, as I discovered when roaming through Jerusalem's underground water infrastructure. The rotating wheels and ultrasonic devices that deliver information about network flow are embedded inside pipes, and there's no central control room where the information they're collecting is fed.

Zohar Yinon, the CEO of Hagihon, Israel's largest water utility (named for the ancient spring), takes us to the bunkerlike basement that once functioned as the utility's control room. It's now just a meeting area with couches and a conference table loaded with trays of cookies. "TaKaDu has put all the controls right here," says Yinon, wagging his iPhone. "I can find out anywhere if my meters are accurate, my water quality is clean, my pressure is good, my flow is normal, my pumps are working properly, my infrastructure is humming. . . . All these layers are integrated online." Hagihon loses about 10 percent of its water supply to leaks and bursts—far less than the 30-plus percent routinely lost in U.S. water networks.

Peleg gives utilities a cloud-enabled service that does more than detect leaks and bursts. It presents the full gamut of information about the water network's operations. Like the software Tony Zhang developed to run his farming operations remotely, Peleg's software uses the Internet of Things to integrate all the layers of information a utility has about its operations into a single interface. TaKaDu's system is used in cities from Sydney, Australia, to Bilbao, Spain, that collectively manage about eighty thousand miles of water pipelines.

The software establishes a baseline of "normal behavior" within each water network. The better it understands normal patterns of water flow throughout the day, the more accurately it can detect aberrations that indicate a leak or burst. It knows that water flows are highest in mornings and evenings, before and after people are at work. It also considers local factors: at a Netherlands utility, for example, the system detected huge and abnormal spikes of flow at regular intervals one Friday afternoon; it noticed that these patterns corresponded with commercial breaks during a World Cup game between Holland and Spain, when fans were flushing toilets. The software can also detect water theft: at Unitywater, a utility in Melbourne, the system noticed abnormally large flows coming from a fire hydrant; officials were notified and they found a strawberry farmer siphoning water from the hydrant.

"Until TaKaDu came along we were basically deaf and blind," says Yinon. "Our network wasn't transparent without this software. It's like an EKG or an X-ray, exposing the inner workings of our system on a real-time basis. We are no longer plumbers and water engineers, we've entered the world of preventive medicine."

Most water utilities worldwide, however, are still deaf and blind, a major concern in an increasingly water-scarce world. Peleg estimates that about 20 percent of utilities globally—and about 10 percent in America—have installed telemetric sensors into their water networks. "Not everybody can see it yet," says Zvi Arom, a member of TaKaDu's board of directors. "There are those utilities you meet and tell them what telemetry and TaKaDu can do, and they say, 'I may as well believe in Snow White and Santa Claus.' "

Water in most regions of the United States has historically been abundant, and most water prices in American cities are low—ludicrously low, says Peleg. The nationwide average, about ten dollars per thousand gallons, is less than half of the prices in Australia and Europe.

Yet the U.S. Geological Survey has predicted that more than three-quarters of the Southwest—a region inhabited by about 65 million people and a significant producer of beans, grapes, onions, potatoes, wheat, barley, and garlic—will be in a state of severe drought by mid-century. Drought patterns have been intensifying throughout Europe, too, along with Russia and China, and the problem has been especially acute throughout eastern Africa and the Middle East. The United Nations predicts that by 2025, Egypt will approach a state of "absolute water crisis." Jordan is one of the driest countries in the world, yet its water demand is expected to double in the next twenty years. Iran is so parched that government officials predict that by 2040, over half of the population will need to be relocated, becoming drought refugees.

Few Middle Eastern nations have begun to import Israeli water tech because of the deep political conflicts between Israel and its neighbors. Yet the harsh realities of failing crops and human thirst have begun to topple those barriers. South Africa is one of many countries that had imposed a ban on Israeli imports in solidarity with its Middle Eastern allies, yet when the drought of 2017 destroyed a third of South Africa's wheat production and reduced the government to rationing water, its leaders agreed to drop the ban and invited Israeli engineers to help rethink its water networks and supply.

California, too, is looking to Israel for answers. Governor Jerry Brown invited Peleg and other water-tech leaders to the state for a summit at which he and Israeli prime minister Benjamin Netanyahu signed an agreement for a blue-tech transfer. Peleg reminded the audience that in July 2015, as California lumbered through its crippling five-year drought, a pipe burst in Los Angeles under Sunset Boulevard, belching 20 million gallons of water. The irony was acute at a time when drought was killing off 200,000 acres of Californian crops. "Our software could have prevented such a burst," Peleg said. "It would have picked up the problem when it was just a small leak."

Peleg noted that, unlike Israel, the United States has no real water policies, no system of carrots and sticks to deter water waste or reward conservation. "In California, they tell me they can't do telemetry because the unions would lose three jobs!" Peleg exclaims. He also stresses the importance of stronger price signals to deter waste. Peleg takes pride in the steep water bills he pays for watering his backyard farm. "Americans think water should be free and unlimited, like air. But the philosophy in Israel is, if you want to have a garden or a pool, fine—pay for it!" Israel has a three-tiered pricing system. "We're only allowed to consume a certain amount of low-cost water for a family of five, for example. Above that quota, the water is 50 percent more expensive. On the next tier, the pricing goes wild."

The pricing structures of U.S. water utilities tend to encourage rampant consumption. "A third of the counties in America still charge a flat rate for water," Peleg observes. "Whether you are a business or resident, you pay a flat rate. Like for $9.99, it's all-you-can-eat water." The policies are as antiquated as the networks, says Peleg, particularly at a time when water-stressed regions like Southern California are beginning to invest in new and extremely high-cost water sources.

Peleg insists future water networks will need to have tiered water pricing and different categories of water quality—"just as at a petrol station, you can choose regular or premium gas. It's lunacy to irrigate crops and flush toilets with the same quality of water you drink in a glass." California recycles about 15 percent of its wastewater, but almost all of the water consumed by the 22 million people of the state's water-stressed southern region is imported, much of it pumped long distances, over mountains, from Northern California. Southern California also draws heavily from the Colorado River, the beleaguered waterway that supplies six other states and Mexico. As these freshwater sources have dwindled both to the north and to the east, the cost of water imported to Southern Californian cities has been climbing nearly 10 percent a year. The changing economics of water have forced Southern Californian utilities to turn in a new direction for relief: westward to the Pacific Ocean.

ISRAEL HAS 120 miles of coastline along the Mediterranean Sea, and California has many times that—840 miles of coastline adjoining the world's largest ocean. It's an oversupply of brine lapping up against an increasingly thirsty landscape. In order to tap this massive reservoir, the San Diego Water Authority partnered with the Israeli company

The hive of desalination cylinders at Carlsbad

IDE to build a $1 billion desalination plant in Carlsbad, a suburb of San Diego. It opened in 2017, the largest desalination facility in the Western Hemisphere.

"If we could ever competitively, at a cheap rate, get freshwater from salt water, that . . . would really dwarf any other scientific accomplishments," President John F. Kennedy told the Washington press corps in the 1960s. Mark Lambert, the head of IDE's U.S. division, who oversaw the building of the Carlsbad plant, describes desalination as "the most significant kind of modern alchemy. About 97 percent of the earth's water is in the ocean, yet only recently have we been able to tap that resource to grow crops or quench human thirst."

In a single decade, desalination took Israel from a water deficit to a water surplus. When drought set in in 2002, the region's already scarce freshwater reserves were squeezed. "We were facing the very real prospect of thirst, but by 2012 we had a surplus," Peleg says. In part it was a victory achieved by conservation technologies and improved recycling, but the problem was bigger than that. Israel needed a new supply, and it would come from a fleet of desalination plants that today produce more than half of the water that flows to Israel's domestic taps.

The behemoth among these plants is Sorek, built by IDE in 2014 and now the world's largest desalination plant, processing 200 million gallons of seawater per day. Located ten miles south of Tel Aviv, the plant is a landscape of concrete, iron, and steel perched at the edge of a quiet beach on the blue-green Mediterranean. Jutting from the sand like a gaping mouth is a pipe six feet in diameter. It slurps in water from an intake out at sea and then belches it into vast concrete reservoirs, where the brine begins its various stages of filtration.

Desalination has been around for millennia if you count the evaporation techniques pioneered by Greek sailors in the fourth century BC. They boiled salt water and then captured the steam. When cooled, steam condenses into distilled water that's free of virtually all contaminants. This same basic technology—thermal desalination—is still used in places like Saudi Arabia, where fuel for boiling the water comes cheap. Since the 1960s, however, most desalination operations use reverse osmosis (RO), a method that simulates the biological process that happens within our cells as fluids flow across semipermeable membranes.

There are still big challenges for desalination. Number one: energy costs. The series of pumps at the Sorek plant collectively exert roughly 7000 horsepower of energy (1100 pounds per square inch of pressure), night and day, to move the water through the membranes. (A NASCAR vehicle does about 700 horsepower at full throttle.) Improvements in the pumps, pipe design, and membranes have cut the total amount of energy used in desalination by about half in the past two decades. It will come down further as efficiencies improve, but many still see the energy demands as a sticking point. Sara Aminzadeh, the executive director of the California Coastkeeper Alliance, one of many environmental groups that have opposed the development of desalination plants in California, tells me, "Desalination may seem like a panacea, but from a cost and energy standpoint it's the worst deal out there."

Helene Schneider, the former mayor of the Southern Californian city of Santa Barbara, decided in 2015 to revive a mothballed desalination plant that was built back in the nineties. Schneider told her constituents that it was "a last resort," but that, given the rising pressures of climate change, they'd have to start tolerating last resorts. This was the mind-set of the San Diego Water Authority when it partnered with IDE to build the Carlsbad plant, which is almost as big as Sorek and provides nearly a tenth of San Diego County's total water supply—enough for about 400,000 county residents. Up the coast from Carlsbad, another large desalination plant is under construction in Huntington Beach, which will supply drinking water to suburbs of Los Angeles. More than a dozen similar plants have been proposed along California's southern and northern coastlines.

But there's another source that's becoming even more critical to the future water supply—one that's harder for some Californians to swallow. The officials call it "recycled wastewater," a pleasant term for human sewage. This is one of the harder realities I've come to accept about third way agriculture—that everything we're now flushing down our toilets and pouring down our drains may have to play an important role in growing our food.

"WE CALL IT the big tooth comb—step one of the filtration process!" Snehal Desai shouts above the sound of sluicing water. There's a visible torrent of raw sewage water flowing through a channel below us at the Orange County Sanitation District, a facility that treats waste from the toilets, showers, sinks, and gutters of 1.5 million suburban Californians. An enormous rake descends into the depths of the sewage flow and brings up cardboard, wet wipes, tampons, egg shells, marbles, toys, tennis balls, sneakers—all the detritus that can't fit through the screen covering the plant's intake.

The flow that passes through the screen has now begun a journey through an advanced purification process that culminates in a stage of RO filtration similar to the process used at the Sorek plant. Daily, the plant pumps out 100 million gallons of drinking water—enough to supply 850,000 county residents—which makes this the largest "toilet to tap" facility on the planet.

The sewage moves through eight stages of filtration, including a gravel-sand filter and a bacterial "bioscrubbing" process used in Israeli plants. Orange County also has a "microfiltration" stage, in which the water is sucked through thousands of tiny porous straws. In the final and most critical stage, the water is forced through a massive hive of cylinders containing the RO membranes.

The solid sewage removed from my drinking water

This Orange County facility is setting a precedent for the use of sewage to produce drinking water every bit as pure as the water that comes from desalination. This process is cheap compared to desalination—about half the cost. Sewage has much lower salinity than seawater, which makes it easier to process. "Recycled wastewater is the fastest-growing area in the water industry—why? Because not every city has an ocean, not everyone has good lakes and rivers, but everybody's got sewage," says Desai. "That's the megatrend."

San Diego recently announced plans to produce 35 percent of its water from recycled sewage by 2030—not just for irrigation but for drinking—and has completed designs on a toilet-to-tap facility larger than Orange County's. Still, there are barriers to overcome. First, the gross factor. Even the desperation of drought can't eliminate the fact that drinking your own waste is nobody's first choice, unless you're a resident of the international space station. Paul Rozin, a social psychologist at the University of Pennsylvania who has consulted water utilities on marketing toilet-to-tap programs to residents, says, "Accepting recycled wastewater is kind of like being asked to wear Hitler's sweater. No matter how many times you clean the sweater, you just can't take the Hitler out of it."

But the purity you get from the RO process is quantifiably better than the water you get from conventional treatments—better even than some bottled water. "What flows from our membranes is the Rolls-Royce of municipal water," says Desai. Whereas tap water is often treated with chemical coagulants and chlorine, RO filtration is a mechanical filtration of water contaminants that cuts the need for those chemicals. It's analogous to the mechanical removal of weeds in a field practiced by organic farmers in lieu of chemical pesticides: "Think of it as 'organic' tap water," says Desai.

For now, Desai is focused on making membrane products for big industrial and municipal water systems, but he envisions micro-scale systems down the line. Bill Gates made a pitch for a similar approach when he blogged a few years back about watching a pile of human feces on a conveyor belt enter a small-scale waste-treatment plant built to serve a community of a few thousand people in Senegal, and, in minutes, get converted into "water as good as any I've had out of a bottle. I would happily drink it every day."

Desai predicts that water filtration technology will become decentralized everywhere—so that we control and regenerate our own water supplies farm by farm, neighborhood by neighborhood, or household by household. Eventually the water production could become, like the food production, circular—a closed-loop system in which 100 percent of water that goes down commercial and residential drains is recycled; whatever is lost in evaporation or leakage can be made up for with desalinated salt water that moves through shared networks. Although the vision is a long way from becoming a reality—decades at least—it may be necessary to our future food security and critical to our survival.

At the end of my tour of the Orange County sewage and water

plants, we arrive at a shining stainless-steel sink. The water that hours earlier had begun as raw sewage was now flowing crystal clear from the tap. Desai filled up two Dixie cups. "To the future!" he toasted. I shuddered as I knocked mine back. But somehow, there was no trace of the Hitler. The stuff tasted every bit as good as water that had bubbled up from a spring in the Alps. I poured myself a second cup.

CHAPTER 11

Desperate Measures

This was the shot that pierced the cloud
And loosed the rain with thunder loud*!*
A shot from the bow so long and strong
And strung with a string, a leather thong;
A bow for the arrow Ki-pat put together
With a slender stick and an eagle feather . . .
A feather that helped to change the weather.

—VERNA AARDEMA, *BRINGING THE RAIN TO KAPITI PLAIN*

IN THE KENYAN folk tale *Bringing the Rain to Kapiti Plain*, a young herdsman named Ki-pat impales a cloud with a homemade arrow to end a punishing drought. The lack of rain has driven wildlife from his region and has begun to starve the cattle that feed his family. It's a story about the essential interdependence of humans and nature, explaining how humans rely on animals who rely on the land which relies on the elements. It's also a story about ingenuity and magic. It presumes that clouds are made of matter that can be pierced, coaxed, and even controlled by engineered tools.

I've read the story of Ki-pat countless times to my kids, and until recently, I'd assumed that attempts at weather control existed only in the realm of ancient rituals and rhyming folk tales. In the summer of 2016,

I stumbled on a short news item about a decision by government officials in Maharashtra, India, to spend millions of dollars to try to force the clouds in their region to rain on the desiccated farmland below.

Maharashtra, which encompasses Mumbai and its surrounding farmlands, is one of the largest and most agriculturally productive of India's thirty states, a major producer of rice, wheat, sorghum, sugarcane, and mangoes. More than 80 percent of its farms depend on rain for irrigation. In 2013, a disturbance in the El Niño storm cycle, which scientists attributed to climate change, reduced the normal monsoon-season rains in Maharashtra by more than half. The drought persisted, and by 2016, food production in the state had dropped by more than a third. The human impact had been severe: tens of thousands of farmers in Maharashtra had committed suicide during the previous three years, unable to grow their crops, feed their families, and dig themselves out of debt.

The public officials of Maharashtra had neither the means nor the infrastructure to desalinate ocean brine or pump filtered sewage water to the state's villages and farms. Maharashtra's water crisis underscores a larger global reality—that even as some wealthy countries are finding ways to conquer drought, many poorer and more populous regions of the world are increasingly threatened by it.

In July 2016, Maharashtra's minister of revenue, Eknath Khadse, took a gamble. He hired Weather Modification, a company based in Fargo, North Dakota, to oversee a multimillion-dollar cloud-seeding program over three years and across sixty square miles of farmland in the middle of the state. Cloud seeding, which has been practiced for decades with varying degrees of success, is a method of injecting a chemical vapor into clouds to stimulate rain. "Our situation is severe," Khadse would later tell me. "There's no other technology available in the world to bring more rains. We must be willing to try it."

Perhaps because Khadse's drought-relief program reminded me of Ki-pat, it caught my attention. I became fixated on the prospect of seeing cloud seeding firsthand. Six months later, I was walking down

Byron Pederson and Shizad Mistry guide us into the storm cloud

the tarmac of a tiny airport in remote Maharashtra to board a King Air B200—the four-seater prop jet that would soon infiltrate a monsoon cloud spanning ten thousand feet from top to bottom, and nearly as wide.

"MOST PILOTS ARE trained to avoid these storm systems," Byron Pederson shouts. "We're trained to enter them." Pederson is the Fargo-based Weather Modification pilot who's leading the cloud-seeding operation in Maharashtra and training a team of Indian pilots to learn the practice. He's flown hundreds of similar missions over the past decade, and I've been told he's as safe and experienced a cloud-seeding pilot as you can find anywhere in the world. This doesn't calm my nerves, however, as Pederson maneuvers the King Air into the sky and rotates the plane sideways, *Top Gun*–style, to circle the cumulonimbus.

The air inside the plane smells rank, like stress and sweat. The view outside my window goes smoky black as we pass through a dark layer of heavy moisture along the cloud's underbelly. The plane lurches and shakes. "We're in," Pederson tells Shizad Mistry, a young Indian pilot in training who's riding next to him in the cockpit. I'm seated a few feet behind them trying not to vomit on the fridge-sized computer next to me that's recording meteorological data. The vertical speed indicator on the dashboard climbs. We've entered the "updraft," a shaft of wind at the center of all storm clouds that's now sucking the plane upward at a rate of hundreds of feet per minute.

"Fire left," instructs Pederson. Mistry flips a switch on the center console and deploys a flare on the left wing. "Fire right." There are twenty-four cylinders resembling sticks of dynamite wired to racks on the plane's wings, twelve on each. The flares are filled with combustible sodium chloride—pulverized table salt mixed with a flammable potassium powder. When the switch is flipped, the end of the flare shoots orange fire and trillions of superfine salt particles are released into the cloud. Water molecules are attracted to salt, so they bond to the particles and grow into raindrops.

Cloud seeding, Pederson tells me, has been practiced in the United States since the 1940s, and in the past two decades has become a fast-growing global industry. Rain-enhancing programs like Maharashtra's are conducted regularly by water utilities in California, for example, to build up the snowpack over the Sierra Nevadas, which melts into reservoirs. The government of China spends hundreds of millions of dollars a year on cloud-seeding operations, often as a means to purge the smog that hovers above its cities. Australia and Thailand are among the many other countries that conduct public and private cloud-seeding operations to enhance their freshwater supply.

The practice, I was surprised to discover, is widely regarded by scientists as environmentally benign. Sodium chloride is nontoxic and has shown negligible impact on ecosystems in cloud-seeded areas. It's also effective—to a point. "There's little dispute that if you can actually get the seeding material inside the clouds, it will enhance precipitation," says Dan Breed, a scientist with the National Center for Atmospheric Research (NCAR). "The question is, by how much, and will the clouds be there when you need them?"

I'd spoken to Breed and other atmospheric scientists before my visit to Maharashtra, and should have been able to deduce then that cloud seeding is not a reliable method for troubleshooting drought—it cannot guarantee food security in water-scarce regions. Pederson himself told me as much before we boarded the King Air. "The hardest part is managing expectations," he says. "People in Maharashtra are hoping for a cure-all to drought. They come out and dance in the streets when it rains, they hug our pilots and say, 'Do it again.' But we can't guarantee that the clouds will be there—and willing to cooperate." Small or wispy clouds can't be forced to rain, he explains; "you need big, moisture-dense clouds for the seeding to work." Even when it works, Breed tells me, the best you can hope for is an increase in precipitation of about 15 percent. That's a lot better than a light drizzle if you're a drought-strapped farmer, but the results are far from guaranteed.

The view from my window in the King Air as the first cylinder fires

It's sheer luck that during our mission over Maharashtra, we have cooperative clouds. Twenty-two minutes after seeding the first cloud, Pederson returns to the location where he fired the initial flare. It's pouring. "We've got drops!" he shouts. He dips the King Air into a victory swoop before gunning over to another cloud cluster. My stomach churns, and I can't hold it in any longer; I vomit into my purse.

It was soon after this incident that I finally understood I'd careened off track—several thousand miles farther off track than I'd gone in Amarillo, Texas, when chasing after cloned cows. Cloud seeding was not, I came to realize, a strategy that could support a secure and equitable food supply. It was a desperate measure in this region, and for many, a false hope. But my trip to Maharashtra, however rash, did offer moments of sobering clarity that eventually pushed me in the right direction.

A few days after our cloud-seeding expedition, I was invited to the home of Honnamma Madivalar, a Maharashtran farmer who, Pederson told me, embodied the dire circumstances that had driven Khadse to pursue the cloud-seeding scheme. Mrs. Madivalar was sitting under a framed photograph of her husband, Ashok, who had committed suicide six months earlier by ingesting a lethal dose of insecticides in his sorghum field. The Madivalar family had accumulated tens of thousands of dollars of debt to a local moneylender over several years to finance seeds, fertilizer, and tractor rentals. Ashok Madivalar had no insurance and no irrigation system, and after three years of failing crops and growing debts, he'd lost hope. In all Indian states, the government gives what's known as a "compensation" of about $30,000 to families if a farmer dies. In the case of Ms. Madivalar and her son, this compensation relieved only a portion of their debt. "That's all his life was worth?" she asks, with a mixture of resignation and outrage.

I understood then what I'd previously known only in abstract terms—that when a food system collapses under the stress of drought or any other pressure, so do the communities that depend on it. India today is still facing the worst water crisis in its history, according to a 2018 study released by the Indian government. More than 200,000 Indians are dying annually due to water shortages. The study predicts that, given climate warming trends, the demand for water in 2030 will be more than twice the country's available supply.

In the last decade alone, nine countries in Africa and the Middle East, including Uganda, Somalia, Kenya, and Ethiopia, have endured crippling droughts. There are now more people living under severe climatic stress than has ever been previously recorded. Throughout the American West, for example, drought intensity has increased 15 to 20 percent in the past two decades. Meanwhile, the low-lying coastal populations of Florida, Louisiana, and Hawaii, for example, are facing the opposite threat—of rising sea levels, pummeling rains, and farm-swallowing floods.

Meeting Honnamma Madivalar made it clear to me that third way agriculture will require a contingency plan. Smart water systems, robotic tractors, vertical farms, and alternative proteins hold promise in wealthy countries, but, for now at least, they're expensive and limited in scope. It was in Madivalar's house that I began to consider more realistic questions—not about how to control the weather above subsistence farms, but about how we'll provide on-the-ground support for populations that are already facing the worst impacts of climate change, including full-blown famine. I began to see what may be the hardest thing to accept about our global food future: we'll have to become proficient in crisis management—and postcrisis resilience.

TWO THOUSAND MILES west of Maharashtra I found my way to the office of Mitiku Kassa, a man who knows more than anyone should about the logistics of famine relief. As Ethiopia's commissioner of disaster risk management, Kassa has confronted every imaginable threat to the country's food security, from droughts and floods to earthquakes, volcanic eruptions, and political upheaval. He's spent his career investigating what to do when a nation's food supply falls short. Kassa's colleagues describe him as unflappable and optimistic, but in the summer of 2015, he was neither. "We were facing the worst emergency in Ethiopia in fifty years," he tells me at his office in Addis Ababa, Ethiopia's capital. "I was worried." The harshest drought in decades had begun to cripple the country's agricultural lowlands, and famine, possibly biblical in scope, loomed.

By August 2015, more than four million Ethiopians were receiving emergency food rations: sacks of wheat, corn, and teff, a grain staple; crates of beans and peas; and jugs of vegetable oil. Soon officials were reporting that these supplies weren't enough. Many low-lying regions of

Ethiopia were far worse off than Maharashtra—they hadn't gotten any rain in nearly a year. The drought had left rivers empty and groundwater overdrawn. Crop yields in these areas were cratering and cattle were dying by the thousands. Acute malnutrition among babies, children, and mothers was on the rise.

In October, Kassa's team calculated that the number of people needing emergency food had doubled in two months, to 8.2 million, prompting the government to request humanitarian aid. By December, 10.2 million people needed food. Kassa was also concerned about continuing to help the many chronically food-insecure Ethiopians who'd been receiving aid even when conditions were stable. All told, Kassa would have to feed more than 18 million people—nearly a fifth of the country's population.

The logistics of rapidly disseminating so much aid in a country almost twice the size of Texas are mind-numbing, as is the challenge of paying for it. "You have to get the dollars in the bank, the food through shipping ports, into warehouses, and into people's homes," says John Aylieff, a United Nations World Food Programme (WFP) executive who has overseen aid programs throughout Africa and Asia and partnered with Kassa on his famine-relief effort. "It can't happen overnight, it might take three months or more."

Kassa didn't have that kind of time. International aid partners, including WFP, UNICEF, and U.S. AID (Agency for International Development), scrambled to fill the funding and supply void, but as they did so, something unprecedented happened: Ethiopia became the lead investor in its own survival. After more than a decade of strong economic growth, the government was able to pump domestic revenue into the drought response—almost $800 million across eighteen months, according to Kassa. The country's aid partners contributed another $700 million. The result was the largest drought-relief effort, with the fewest human fatalities relative to the scale of the crisis, that the world had ever seen.

The success of the effort wasn't only or even primarily about cash flow—it stemmed from the many years, even decades of preparations the country had already made. In a sense, the curse of past traumas became the blessing: Ethiopia had learned to live in a constant state of anticipation and preparation—a state of readiness that many other drought-prone countries may soon have to learn to adopt. Mary Robinson, the U.N. special envoy on climate change, expects this will be the case—that other nations in and beyond southwestern Africa will look to Ethiopia "for a blueprint in building resilience against the climate pressures ahead."

KASSA IS FIFTY-TWO and a broad-shouldered, barrel-chested six foot four—a man who, you can tell at first glance, has never experienced severe hunger. He grew up in Ethiopia's lush southwest, working on his parents' prosperous farm, which grew coffee, citrus, and guava trees. He was studying agriculture at Haramaya University during the drought-fueled famine of 1984, which killed nearly a million people and left many more destitute. The famine was heavily televised around the world, including in most U.S. households, and is the reason, Kassa says, "Ethiopia has become synonymous in people's minds with 'poverty' and 'famine'—an association that is a source of considerable shame for our people." Although Kassa was personally insulated from the tragedy, which struck regions of the country that were drier and more vulnerable than his own, it haunted him. He received a scholarship to Wageningen University in the Netherlands, where he again studied agriculture, then returned to Ethiopia in 1998 to become head of the country's Planning and Economic Development Department, which oversaw agricultural innovation. "He's a remarkable person—preternaturally composed," says John Aylieff. "I've never seen him get frustrated, even in the toughest times. He is always just a bottle of calm."

Ethiopia has faced multiple droughts since 1984, including severe ones in 2000 and 2011, after which Kassa established a network of roads, warehouses, and local distribution points for food aid. This infrastructure, he says, is the single most important line of defense

Mitiku Kassa

against famine—without it, you can't distribute relief. He also established a monitoring system to regularly assess farm productivity and the scope of nutritional needs from region to region—data that's crucial to anticipating and responding to potential disruptions in food supply.

When the first symptoms of drought and food scarcity were reported to Kassa in 2015, he began taking weekly trips, often traveling thousands of miles to explore impacts in the worst-hit regions. Kassa, who is a devout Christian, calls the travel a practical necessity and an exercise in empathy. "A love for humanity isn't given by training," he tells me. "It is given by life experience and the almighty God."

Ethiopia is diverse—its people speak about ninety different languages and its topology and climate zones vary significantly, so much so that many large agricultural regions were untouched by the drought. Kassa knew that while some aspects of the relief program would be universal, others would need to be tailored. The northern Afar region, for example, was suffering from particularly severe water scarcity, so he assigned 132 trucks to continually deliver drinking water to the local distribution points there. The eastern Somali region, which is home to almost three-quarters of the country's livestock, needed water pumps for fodder-supply programs and vaccinations for disease-prone animals.

Amhara, in the northwest, had untapped groundwater resources, so it required funds and equipment for well drilling.

Kassa and Aylieff described nerve-racking periods when critical supply shipments were held up in Djibouti, or when aid money hadn't come through. "There were many nights we were up past midnight, when our warehouses were down to their last sacks of wheat or our accounts were down to their last dollar, and we were on the phone pleading for shipments, begging for donor support," says Aylieff.

The effort benefited, in a twist, from Ethiopia's powerful central government, which had been widely criticized as too powerful, even dictatorial. Civilian protests in 2017 railed against that oppressive leadership. But it was in part because of the government's unmitigated power that its leaders could move quickly to secure funds and mobilize relief as famine approached. In the past decade, Ethiopia had one of the fastest-growing economies in Africa, with its textile and agricultural sectors booming, along with its cities and road network; agricultural productivity has doubled in the past thirty years. The government, however corrupt, now had wealth to tap—and the power to tap it. "It's like the Chinese government," one U.N. official told me, "very responsive and decisive at a time of economic or environmental crisis, but brutal in their democratic leadership." The lesson to be taken from this, the official stressed, is not that corrupt governments will be the most effective at feeding their vulnerable populations, but that democratic governments should establish emergency funds and protocols for the swift and decisive deployment of these funds if disaster strikes—"because when people are hungry, speed is everything."

Within a year of the start of this drought, Kassa and his partners had hauled more than 12 billion gallons of drinking water, sunk hundreds of bore wells up to fifteen hundred feet deep throughout the country, and sourced and distributed almost 2 million metric tons of maize, wheat, teff, legumes, peas, beans, and palm and vegetable oils. Some of these crops came from government-owned farms in Ethiopia and neighboring countries; others came from places as distant as Ukraine,

Canada, the United States, and Australia. The relief effort between 2015 and 2017 was so swift and comprehensive, says Kassa, that there was "no loss of human life." John Aylieff of the WFP calls it "the world's first famine without human casualties."

Not everyone agrees with this claim. Some news coverage of the famine questioned government data and reported a rise in Ethiopia's mortality rates during 2016 and 2017. But Gillian Mellsop, UNICEF's top representative to Ethiopia, who compiled and assessed data from hundreds of health posts in drought-afflicted regions, says the most striking thing about Ethiopia's recent famine is that "the data showed no notable increase in infant or child mortality rates, even though there were more than 2.5 million children diagnosed as severely malnourished." Child mortality rates are the most important data to track during a famine because the under-5 population is the most vulnerable to nutrient deficiencies.

"If there was a failure in Kassa's approach," says Alemu Manni, a field coordinator at the U.N. Food and Agriculture Organization (FAO), "it was that we need to invest more in the kind of strategies that will, down the line, keep farms up and running and eliminate the need for food handouts. Communities need their own bore wells, their own mechanized equipment, their own seeds—that's where the majority of funding should be going." This makes economic sense: the cost of food rations to sustain a family is about twenty times higher than seed provisions. Similarly, the cost of vaccinating livestock and creating fodder-supply programs is far less expensive than the cost of restocking deceased animals. "At a certain point, food aid won't be enough," Manni warns. "The only communities that will survive these droughts will be those with built-in resilience."

Kassa did manage to use a portion of the budget at his disposal—if a small portion—to help implement some innovative programs. For example, he established a new food distribution program through the country's elementary schools. Famines tend to cause a sharp increase in school dropout rates, because hungry kids burn energy walking to and from school and their families may need them to help find food. So

Kassa distributed aid through the schools themselves, offering vitamin-fortified cereals at lunch with a selection of vegetables, fruit, grains, and milk—meals with more nutritional diversity than many children could get at home.

In partnership with the FAO, he supported programs such as pop-up poultry cooperatives with mobile chicken coops, to supply eggs to regions with limited protein sources. In livestock-dependent regions where water pumps couldn't be installed to grow fodder, the FAO taught herders how to make and sell "multinutrient blocks" for cattle—high-calorie nutritional supplements made of grains and molasses.

In 2018, Ethiopian farmers reaped their first healthy harvest in two years, and agricultural yields increased 20 percent nationwide. Yet Ethiopia's neighbors Somalia, Kenya, and Uganda were still facing devastating water scarcity. For Kassa, there was no relief—it was time to start preparing for the next drought.

A dried-up riverbed in Ethiopia

On a late-summer day, Manni takes me to the Somali region in eastern Ethiopia to show me what built-in resilience looks like. We travel through pastoral lowlands that are sun-scorched for miles in every direction; the topsoil is gray and cracked like old paint. The herds of cows we pass are listless and gaunt, their hides sagging between their ribs as they scrounge for grass. Hunger had killed thousands of cattle in this region in 2016. But at the edge of an empty riverbed, an oasis suddenly appears: seventy acres of elephant grass and Sudan grass, papaya and mango trees, maize, sorghum, peanuts, peppers, and cabbage.

Hamdi Muhammed Mowlid, a twenty-eight-year-old father of four, carried bundles of grasses to cows and oxen lounging in the shade. "*Bozzänä!*" he shouts at the animals. The word translates roughly from the Amharic as "couch potatoes," and he means it as high praise. Mowlid comes from a long line of cattle herders, people who migrated hundreds of miles with their animals each season in search of edible pasture. Now he's discovering the advantages of staying put. Lazy cattle burn fewer calories and are more productive—each of Mowlid's cows provides as much milk as ten of his father's. He sells his surplus, about five liters daily, to a local dairy, generating profits for his family of about thirty dollars per week. The sales are coordinated by the Hodan Fodder Cooperative, which Mowlid helped establish alongside twenty-five neighboring families in 2015. Hodan was a small but hopeful success among the community-based resilience programs that Kassa and the FAO were able to fund during the crisis.

At the edge of the oasis, a pump the size of a lawn mower engine draws water from deep below the riverbed into a network of hand-dug irrigation canals. During the drought, the water pumped by this single engine, which is no bigger than a shoe box, sustained crops that fed 200 people and provided forage for 150 cattle. "We have fattened our cows, fattened our children, and sold our surplus," Mowlid tells me. "We learned what our fathers didn't know, what our grandfathers didn't know." The cooperative also runs a kind of cow-flipping operation, buying skinny cows at a price of about $150 each, fattening them on fodder for six months, and selling them for $300. "In drought," he says, "we have found a better way to live."

CHAPTER 12

Antiquity Now

Water, is taught by thirst;
Land, by the oceans passed;
Transport, by throe;
Peace, by its battles told;
Love, by memorial mould;
Birds, by the snow.

—EMILY DICKINSON

FEDERAL HIGHWAY 200 snakes some fourteen hundred miles down the Pacific coast of Mexico, past volcanoes, craggy mountains, pitchfork cacti, and cattle ranches with skulls on their barbed-wire fences. Somewhere between Tepic and Tapachula, the road reaches Agua Caliente, a town named for its hot springs, which bubble into a stream at the center of a deep valley. On the western margin of Agua Caliente, Mark Olson, a professor of evolutionary biology at the National Autonomous University of Mexico, has a farm. "It may look like a shitty little field with runty little trees in a random little town, but it's an amazing scientific resource," Olson says as he leads me through the hilly, hardscrabble acre that constitutes the International Moringa Germplasm Collection. This is the world's largest and most diverse aggregate of trees from the genus *Moringa,* which, Olson believes, are "uniquely suited to feeding poor and undernourished populations of the dryland tropics, especially in the era of climate change."

These dry tropical regions span most of India and large portions of sub-Saharan Africa and Central and South America. More than two billion people, about a third of the global population, live in dry tropical zones. According to David Lobell of Stanford's Center on Food Security, these regions are vulnerable to the worst impacts of climate change. "If you look at climate models, the conditions are projected to intensify here more than in most other climatic regions. So the already hot, dry climates will become really hot and really dry, relative to their current state," he says.

Hot and dry are precisely the conditions in which the *Moringa* thrives. "This is an ancient plant so tenacious, resilient, versatile, generous, and flat-out eccentric as to be Dr. Seussian," Olson tells me. "Nothing else in the plant kingdom really compares." Although the *Moringa* is neither striped nor candy-colored, it does bear a certain resemblance to the truffula tree, with its smooth, skinny trunk and affably chaotic branches, which protrude like hands waving hello. Not only does it succeed in harsh conditions, it also grows weed-fast—about a foot per month, to a height of as much as twenty feet. *Moringa oleifera*, the most commonly farmed species, is a nutritional Swiss army knife: it produces edible leaves that are unusually rich in protein, iron, calcium, nine essential amino acids, and vitamins A, B, and C. Its seedpods, which are about the width of a thumb and more than a foot long, are also high in omega-3 fatty acids. Jed Fahey, a biochemist at Johns Hopkins University's Bloomberg School of Public Health who has collaborated with Olson on *Moringa* research for more than a decade, has found that the tree's leaves and pods have strong anti-inflammatory and antidiabetic properties, and contain enzymes that protect against cancer. Mature *Moringa* seeds can be pressed for vegetable oil, and the seedcake that is left over can be used to purify drinking water. (It contains a protein that makes bacteria stick together and die.) When dried, the crushed seeds can also serve as a good fertilizer.

While the *Moringa* tree is an anomaly, it's one among many hardy, ancient foods and crops that may be poised to make a twenty-first-century comeback. These foods range from familiar grains such as quinoa and amaranth to more eccentric plants like *Moringa* and Kernza (a type of wheatgrass), and even to microvegetation, like algae and duckweed, which you may not consider edible at all.

Olson contends that global warming and population trends are forcing us to think differently about the quality and resilience of the crops we grow, both in the poorest parts of the world and the wealthiest. It's no longer enough to do high-volume agriculture—future crops must be higher-*quality*, meaning more nutritious while also hardy, fast-growing, and tolerant of climate volatility. "To find these crops, we have to explore our botanical past," says Olson. Does he consider himself a kind of ancient plant whisperer? I ask. No, he answers, "I'm more of a plant *listener*." Plants like *Moringa* spent millennia learning to adapt to extreme and harsh conditions without modern irrigation, fertilizers, and pesticides—"it's really the plants that have the wisdom. We scientists have to find a modicum of humility and look to them for the answers."

MARK OLSON GREW up on the edge of Tahoe National Forest, in California, where his father worked as a civil engineer for the U.S. Forest Service. He spent his childhood catching newts and lizards, collecting insects, and raising orphaned baby birds. In junior high he worked as a volunteer in a zoo and a wildlife hospital, but it wasn't until college at the University of California, Santa Barbara, that he got deep into botanical research. "I paid for most of my education working at the herbarium of the university," he tells me. (I then googled "herbarium," which as it turns out is a collection of dried plants.) He went on to pursue his graduate degree in botanical research in the tropical dry forests of Mexico, which, he observes, are "the habitats on earth with the broadest variation in plant life-forms—the birthplace of the great icons of the plant kingdom, from giant 'bottle trees' like baobabs, flat-topped acacias, to a great profusion of tree succulents."

Olson talks about botany the way a (possibly tipsy) sommelier talks about wine. He has pale blue eyes and wispy, sand-colored hair, and he wears round metal-rimmed glasses and a leather cowboy hat that make him look equal parts Teddy Roosevelt and Crocodile Dundee. He's been studying *Moringa* since 1995, before concerns about climate change had entered the mainstream consciousness. With funding from the National Science Foundation and the National Geographic Society, he spent nearly two decades collecting the seeds of the tree's thirteen known species, traveling throughout Southeast Asia, the Middle East, and eastern Africa—along the way, listening as carefully as he could to whatever ancient wisdom the plants had to offer. To deepen his research, Olson actually hand-built a personal strap-on helicopter he dubbed a "paramotor," composed of a backpack, a large propeller, and a two-stroke engine.

"The paramotor makes you into like a human hummingbird," he says, "so it's possible to stay aloft for hours close to the ground and explore the crowns of trees." He piloted the thing with hilariously mixed results. On an early flight, he stalled the engine and came crashing to the ground, shredding his propeller. Olson thereafter managed to study hundreds of *Moringa* trees with his paramotor without losing a limb, but his focus now is on more sophisticated tools that he hopes will help

Mark Olson prepares for takeoff in his paramotor

make *Moringa oleifera* a staple nutrient source for billions of people in climate-stressed regions.

There are a number of limitations *Moringa oleifera* needs to overcome. The broad genetic variation within its seeds makes it hard to grow on a large scale because uniformity in row crops is critical for ease of care and harvest. Its leaves are smaller and more delicate than baby spinach, and are prone to wilting after they're picked—a challenge for farmers who can't chill their produce. *Moringa*'s limitations are culinary, too. Its leaves, like coriander, taste best when removed from their chewy stems, a tedious process when cooking large quantities. Both the leaves and the pods contain an oil that gives them a pungent, peppery flavor—like rocket, but stronger—which can be off-putting. In India, where *Moringa* was first domesticated two thousand years ago, the pods are commonly used in a popular dish called sambhar, which subdues their flavor in a rich gravy. During my visit to Agua Caliente, Olson prepared tacos with parboiled *Moringa* leaves, *queso blanco*, and salsa, which were delicious, but the raw leaves, straight from the tree, were bitter and hard to swallow.

Olson's research on his farm in Agua Caliente is focused on adapting *Moringa* to modern palates and production needs. He wants to develop a *Moringa* tree that has subtler flavors and is easier to harvest, but still hardy and nutritious enough to become a staple food source in dry tropical regions all over the world in the coming decades. He's also working with scientists in India to sequence the *Moringa oleifera* genome, with the hope that genetic breeding tools can help pave the way for this optimal breed—a kind of "killer app" of ancient plants.

FOR NOW, *MORINGA* is gaining more popularity among Western superfood enthusiasts than among the underserved populations of the dry tropics. "*Moringa* is the new kale," says Lisa Curtis, the thirty-year-old founder of Kuli Kuli, a San Francisco–based company that specializes in *Moringa* snack bars, powders, and energy shots. The Kuli Kuli website, which refers to *Moringa* alternately as a "miracle tree" and a "supergreen," states that the plant roundly outperforms kale, with "2 x protein, 4 x calcium, 6 x iron, 1.5 x fiber, 97 x vitamin B_{12}." (Olson said that he hasn't seen data to support these nutritional claims but that the nutrient profile of *Moringa oleifera* "does at least rival or exceed that of milk, yogurt, and eggs, serving for serving.") Whole Foods now carries *Moringa* products nationwide and in 2017 named Kuli Kuli the Whole Foods "Supplier of the Year." Curtis sold more than $3 million of *Moringa* products in the first six months of 2018.

Olson was skeptical at first of the *Moringa* fad within wealthy enclaves. "Trumpeting dried *Moringa* as the cure du jour for people in the rich West misses the real potential of this plant," he says. He sees *Moringa* as a kind of anti-superfood—not something to be frittered away as a luxury supplement, like açaí berries sprinkled on oatmeal, but to be used as a staple form of sustenance. For now, though, *Moringa* is a hard sell in many parts of the world, including in the very town where Olson has his International Moringa Germplasm Collection. When I asked a local farmer in Agua Caliente whether his family eats *Moringa*, which grows in the yards of many houses around town, he replied, "*Es ayuda contra el hambre*"—it's a famine food, a last resort.

Olson is recruiting chefs in Agua Caliente, where he spends several months a year, to develop *Moringa* recipes and educate their neighbors on the tree's nutritional value, but as Curtis sees it, the trend among wealthy American consumers can help spur demand for the crop throughout the dry tropics. Curtis sources her products from dozens of farming cooperatives in India, eastern Africa, and Central America. "The farmers get it," she tells me. "They know that Americans want *Moringa* and that elevates its value. They see that it's not a famine food but nutritional gold."

Moringa is one among a cohort of ancient plants that possess both climate resilience and nutritional gold. In the United Kingdom, Sayed Azam-Ali, director of the University of Nottingham's Crops for the Future Research Centre, says that "feeding the world sustainably and nutritiously may require bringing back online dozens of ancient crops you've never heard of." Azam-Ali describes a range of ancient plants that possess properties uniquely suited to solving modern problems. Among those he and his colleagues are researching: a fava bean ancestor that's protein- and potassium-rich and also good at fixing soil nitrogen; Bambara groundnuts—a highly drought-resistant relative of the peanut; and, further afield, *Wolffia globosa,* an edible species of duckweed that has a similar protein content to soybeans and peas.

The mung bean is another ancient plant with new relevance today. The 4300-year-old legume from Asia contains proteins strikingly similar to egg proteins—so much so that the San Francisco–based food company JUST has made it the foundation of a new nonanimal, zero-cholesterol scrambled egg substitute that has been praised by chefs and executives alike: "It's not every day you see something that blows your mind," said celebrity chef José Andrés.

Quinoa, too, is an indigenous, protein-rich superfood that has entered a modern renaissance. Quinoa was first domesticated about seven thousand years ago around Lake Titicaca in the Andes and became known as the "mother grain" of the Inca empire. Today production is still mostly limited to the Andes region. Bolivia and Peru produce more than 90 percent of the world's quinoa and have protected this legacy by restricting the release of their heirloom seeds to international growers. But rocketing demand on top of limited supply has tripled the cost of quinoa in the last decade, prompting scientists and entrepreneurs in the United States and Canada to try their hand at breeding and growing the crop.

One such entrepreneur is David Friedberg, a former Google programmer who has become an investor in food and farming technology. He cofounded a San Francisco–based fast-food chain, Eatsa, serving only bowls of quinoa with various toppings, and owns the company NorQuin, which is now run by former Wise company CEO Aaron Jackson. NorQuin produces tens of thousands of acres of quinoa on a farm in Saskatchewan. "Quinoa is a finicky crop that likes a certain amount of cold, and a particular type of dry, saline soil," Friedberg tells me. "There's huge potential to adapt it to warmer and more fertile growing conditions." Friedberg and Jackson have identified more than 45 million acres throughout Canada and the United States that they think will, with advances in breeding, be suitable for quinoa production. Eventually they see quinoa becoming competitive with rice, wheat, and soy as a primary food for the climate era. "If you were to start the human civilization today you wouldn't say, 'Let's grow rice and wheat as our staple crops,'" says Friedberg. "They require an incredible amount of water and have subpar nutritional value."

An agronomist named Wes Jackson is another key figure in the effort to bring ancient plants to modern consumers. Jackson is the founder of the Land Institute, a Kansas-based research center for sustainable agriculture, where his team has spent the past decade cultivating Kernza. This breed of wheat is a blend of old and new—derived from wild perennial wheatgrass that's been growing on the Kansas prairie for millennia. Kernza produces grain for up to five years and grows ten-foot roots that can tap deep reserves of groundwater. Compare that with conventional wheat, which produces grain for one year only and grows roots that are less than half the length. Kernza also channels and locks carbon dioxide into the soil while also enriching its health; if grown on thousands of acres, Kernza farms could sequester vast quantities of greenhouse

Agronomist Jerry Glover shows off Kernza's climate-resilient roots

gases. The grain requires far less tillage, fertilizer, and water than conventional wheat and protects against erosion. First adopted by boutique brewers and bakers on the West Coast, Kernza has attracted the attention of General Mills' Cascadian Farms, which is now incorporating the grain into its cereals, snack bars, and crackers.

Another growing area of research: edible algae. The algae known as spirulina has been eaten in some parts of the world as a protein supplement for aeons, but, like *Moringa*, it has a strong flavor and limited appeal. Now, using advanced breeding techniques, scientists have developed strains of edible algae that are flavorless and odorless and could become an additive in foods—filling the role that soy powders have played for decades—as a protein substitute or filler in meat products, cereals, and breads. Like *Moringa*, algae could be a source of plant-based protein for companies like Beyond Meats, Impossible Foods, and others in the booming synthetic meats industry.

Rose Wang, the founder of the company Chirps Chips, is also making a push for another traditional form of nourishment: edible insects. Insect farming has been gaining ground for years as a substitute for soy in animal feed, but Wang sees it becoming a staple of human diets. Chirps Chips produces a line of snack foods made from powdered crickets, and Wang has managed to get her products into 1500 stores, including Kroger supermarkets, Vitamin Shoppes, and cafeterias at Disney World. "The first time I was dared to eat a scorpion in China, I was terrified. And the moment I bit down, all I could think was *Wow! This tastes like shrimp,*" she tells me. "That's when the future opened up."

Wang and her partners are introducing other cricket-based snack foods, including Chirps cookie mix, into their product lineup, but they argue that the biggest potential role for insect proteins is replacing, as edible algae might, the soy powders that are used as fillers in processed food products. "I see insect powders becoming a dominant protein substitute," Wang says. "Fifty percent of millennials want to eat less meat and are looking to alternative protein sources. Insects use a thousand times less water and produce one percent of the emissions of cows. They also have more protein and a third the fat of beef. The biggest hurdle is getting people over the 'ick' factor. And that's where we come in."

THE OBSESSION AMONG wealthy consumers in the United States with high-nutrient superfoods has grown in part because of the inverse trend—a broader decline of nutrition in mainstream diets. "Many of our most important foods have become less nutritious over time," says Jo Robinson, a plant historian and the author of *Eating on the Wild Side*.

Robinson explains that humans have been evolving for millennia to crave food that isn't good for us; in fact, we've been trending toward "bad"—pleasure- and convenience-driven—eating habits since the dawn of agriculture. Remember that farming itself dealt a nutritional blow to early humans: in exchange for the convenience of stationary food production, the first settlers incurred widespread stunting and diseases related to nutrient deficiencies. Since then, farmers have been selecting for plants that are relatively high in sugar, starch, and oil while also comparatively low in beneficial things like fiber and antioxidants.

The trend has been that "the more palatable our fruits and vegetables became, the less advantageous they were for our health," says Robinson. "Many of the most beneficial minerals and phytonutrients in food have a sour, acrid, or astringent taste that we selected against." This explains why *Moringa oleifera*, still genetically similar to its ancient predecessors, has a bitter flavor. Today, the foods that most closely resemble their wild ancestors, like rocket, Brussels sprouts, and most herbs, generally have high nutrient value but also strong flavors, and we consume them in relatively small amounts.

Farmers and agronomists haven't been dumbing down nutrition intentionally—in fact, as Jo Robinson observes, "only recently have we had the tools to analyze the nutrient content of plants in a detailed way." She cites examples of ancient soybeans, for example, that have about five times more omega-3s than modern varieties; of wild dandelion greens, a favorite of many Native American tribes, that have seven times more phytonutrients than spinach; of a purple potato native to Peru with nearly thirtyfold more flavonoids than russet potatoes; of an apple that has one hundred times more phytonutrients than the Golden Delicious. While many of us have come to accept that flavor and nutrition in the mass-produced crops we eat have declined over time owing to soil quality, long-distance food distribution, and other factors, a more concerning reality is that climate change may worsen this trend.

In 2002, Irakli Loladze, a graduate student at Arizona State University, began to study a phenomenon later dubbed the "junk food effect." He noticed that elevated CO_2 in the atmosphere accelerates photosynthesis, the process that helps edible plants convert sunlight into food. This would seem to be beneficial—causing the plants in his experiment to grow faster. It did that, but it also caused them to accumulate more carbohydrates like glucose—leaving less room for other crucial nutrients and minerals to develop, like protein. Loladze has since observed this process happening in the field, even in wild plants and weeds: "Every leaf and every grass blade on earth makes more and more sugars as CO_2 levels keep rising," he told *Politico*. "We are witnessing the greatest injection of carbohydrates into the biosphere in human history."

In 2017 and 2018, the Harvard University School of Public Health published studies bearing out this trend in more detail. The upshot: staple future crops will be lower in protein and minerals due to rising levels of carbon dioxide. A team of scientists led by Professor Sam Myers analyzed data from field-study tests performed at sites in the United States, Australia, and Japan. The tests examined dozens of different food crops under the atmospheric CO_2 levels we are likely to experience forty years from now. The scientists found that in grains such as wheat and rice that sustain vast populations, zinc, iron, and protein declined 3 to 17 percent.

The junk food effect makes the AeroFarms approach seem all the more appealing: CO_2 levels can be controlled within indoor vertical farms, and nutrient levels in our fruits and vegetables can be finely calibrated. This method, though, raises the cost of fresh produce (at least in the near term), making nutritious foods even less accessible to the populations that need them most. Myers sees three other possible ways to adapt. "One, we could engineer new strains of crops that aren't as sensitive to atmospheric carbon dioxide levels," he says. "Two, we could breed crops with higher nutritional content to begin with. Three, we can increase the amount of fruits and vegetables in our diets to offset the losses."

Option three would likely be the hardest to pull off. Increasing the consumption of fruits and vegetables is unrealistic in low-income countries with scarce arable land and booming populations. It's unrealistic, for that matter, in the United States. According to a recent study by the Centers for Disease Control and Prevention, only one in ten Americans currently eats the daily recommended amount of fresh fruits and vegetables. About 65 percent of the U.S. population eats less than two servings of fresh produce a day. The lowest consumption is among populations that have no access to fresh food markets in their neighborhoods. Nearly 25 million Americans live in so-called food deserts, or neighborhoods with limited access to healthy and affordable foods, and more than half of this population is below the poverty line. "We've built a food system in which poor people are much more likely to eat unhealthy foods," says Hilary Seligman, a professor of nutrition and food policy at the University of California, San Francisco. The income gap, in other words, has led to a nutrition gap, and to fill that gap we need to make fresh and nutritious foods affordable, accessible, and desirable to people across the income divide.

Yet these goals feel out of reach, especially given the other, more literal, junk food effect that's been plaguing our food system in recent decades. The mass marketing of processed foods high in salt, sugar, and fat has encouraged diets that are rich in calories but low in nutrients. Average sugar consumption in the United States has jumped more than 30 percent in three decades; during that time, the weight of the average American adult has increased about 20 percent. The prevalence of diabetes has increased 700 percent.

"We're naïve to think that malnutrition is occurring only in poor or developing countries," Seligman tells me. There are, in a sense, two different types of nutritional famines occurring worldwide: one among climate-stressed populations with too little food, the other among industrialized populations with too much of it. This paradox underscores the truth in Mark Olson's assertion that food producers must shift their focus away from quantity alone. "Feeding humanity in the coming decades may be less a challenge of growing higher quantities of food than of growing higher-*quality* foods," he says.

IT'S TEMPTING TO look out twenty or thirty years and imagine utopian scenarios of quinoa and Kernza replacing rice and wheat fields throughout the world by the millions of acres; of sustainable algae farms replacing environmentally damaging soy farms; of *Moringa* farms providing a secure nutrient supply throughout the dryland tropics. And while these ancient superfoods certainly have the potential to help restore nutritional integrity to our future food supply, they may not succeed on a large scale.

Kernza, which is still in the very early stages of breeding, has grains that are less than a quarter the size of conventional wheat grains, making the harvests tedious and costly by comparison. Using traditional breeding methods, it could take twenty years or more for scientists to develop a version of the plant that can be grown on a mass scale. The market for quinoa, meanwhile, is still a fraction of the size of the market for rice in North America, and it could take at least another decade to get the cost of production down to thirty cents a pound (competitive with rice) and still longer to get the quality and flavor up to par with Bolivia's and Peru's products.

Moringa currently has an even smaller market than quinoa, and Mark Olson predicts it could take him "the rest of my life, at least" to breed the "killer app" of *Moringa* he dreams of—unless he's able to use gene-editing tools like CRISPR to speed up the breeding process. Olson is working with a team of scientists in India to sequence the *Moringa* genome with this in mind. Once the genome is sequenced they can isolate, for example, the genes that code for the plant's abundant production of proteins, iron, calcium, and vitamin B_{12}. "There's little doubt that ancient plants combined with modern breeding tools could more rapidly close the nutrition gap," says Olson.

Olson harvests Moringa

In the long run, neither Olson nor Lisa Curtis of Kuli Kuli envisions *Moringa* becoming a major industrial row crop like maize or soy. Instead, they see it succeeding on single-family and community scales. But a genetically optimized *Moringa* tree would enable these small-scale farmers to produce higher-volume, higher-quality, milder-tasting crops that are more uniform. If the *Moringa*'s flavor can be subdued and its production scaled up, Curtis sees it becoming an ingredient not just in snack bars and smoothie powders, but as a wholesome additive in popular processed foods, like burgers and crisps.

Olson is focused solely on supporting *Moringa* as a staple food for subsistence farmers. Once he's finished his breeding research, he plans to develop a program that will distribute seeds for communities to grow "protein plots," with about twenty trees per family, throughout the dryland tropics. Over time, this could establish a drought-resilient food supply for the populations that need it most. "When you look at maps of the areas in the world where *Moringa* grows, and then at maps where populations are undernourished, it's amazing—they almost exactly overlap," Jedd Fahey of the Bloomberg School of Public Health tells me. This correlation may strengthen in the coming decades—as these climates become hotter and drier, and as the *Moringa* is bred to withstand ever harsher conditions.

As for quinoa, a team led by Mark Tester, a plant scientist at King Abdullah University of Science and Technology in Saudi Arabia, sequenced the genome in 2017. "This crop is a nutritional and ecological marvel," says Tester. "The goal was to get it out of the health-food section and into the realm of a global commodity." The genome sequencing "has put the breeding process on steroids," creating opportunities to rapidly produce higher-yield varieties adapted to different growing regions.

Tester's team has identified genes in quinoa that currently limit its quality and productivity. One, for example, produces saponin, a bitter toxin that the plant excretes on the skins of its seeds and flowers to repel insects and birds. Low-saponin quinoa would ratchet down production costs. Tester also hopes to breed quinoa with shorter stalks, higher yields, and adaptability to different growing regions.

Quinoa's unique ability to tolerate water scarcity and salty soil and to produce protein-rich grains could be a beneficial characteristic to share with crops like rice and barley. In theory, the opportunity scientists now have is not just to breed nutritious superfoods for a mass market, but also to breed mass-market crops that are imbued with the nutrition levels closer to those of superfoods.

Golden Rice has been one fraught but well-intentioned example of this. Other lesser-known efforts to address severe region-specific nutrient deficiencies may hold more promise. Nigerian scientists have developed GMO cassava enriched with vitamin A. Indian scientists have engineered pearl millet with high levels of iron and zinc. HarvestPlus, an NGO backed by the Gates Foundation, is working to produce zinc-enriched rice, lentils, and wheat for global consumption while also testing iron-fortified beans in Rwanda. James Dale, director of the Centre for Tropical Crops and Biocommodities at Queensland University of Technology in Australia, has been trying to engineer "super bananas" spliced with genes from a rare plant in Papua New Guinea that gives them many times the beta-carotene levels of conventional bananas.

While critics of genetic engineering and gene editing bristle at these efforts, they're an extension of so-called biofortification programs that have been going on for nearly a century. In 1924, salt imbued with iodine ("iodized salt") was introduced in the United States as the world's first nutrient-fortified product. The goal was to quell an epidemic of goiter, which is caused by iodine deficiencies, and the effort succeeded. Milk fortified with vitamin D emerged in the 1930s to combat rickets—also successful. The following decade brought flour enriched with iron, thiamine, and folic acid to prevent anemia, which contributes to one in five maternal deaths worldwide. In the 1980s, beverage makers began to fortify water, juices, and soft drinks with calcium to address child stunting and osteoporosis.

Critics see the GMO fortification trend as a misguided effort to slap technology onto problems wrought by technology. Going forward, it may be possible not just to improve nutritional quality of staple crops, but also to restore ancient traits of climate resilience. Tester is working to isolate the genes in quinoa that enable it to grow in salty soil—a trait that could be valuable if transferred to staple crops grown in regions like coastal India and Bangladesh that are prone to saltwater flooding and rising sea levels. Olson, meanwhile, is eager to isolate the genes that govern what he calls *Moringa*'s "ingenious plumbing." The tree, it turns out, has unique mechanisms for drawing the water up from the earth, through its roots and then through special tapered channels in its trunk into its leaves. Its special ability to harness and store water may be, perhaps, the tree's most valuable and Dr. Seussian characteristic—the one that most makes it, like the truffula, an ecological linchpin. Even if the *Moringa* does not become widely adopted as a food source, it may help agronomists understand how other food-bearing trees will and could behave in increasingly water-scarce conditions.

In the end, Olson's research on the *Moringa* tree is, like Tester's quinoa research and Wes Jackson's Kernza research, profoundly optimistic. It reminds us that the wisdom of the past, as much as the ingenuity of the present, can help us survive the future.

CHAPTER 13

What Rough Feast

A human being is primarily a bag for putting food into; the other functions and faculties may be more godlike, but in point of time they come afterwards. A man dies and is buried, and all his words and actions are forgotten, but the food he has eaten lives after him in the sound or rotten bones of his children. I think it could be plausibly argued that changes of diet are more important than changes of dynasty or even of religion. . . . Yet it is curious how seldom the all-importance of food is recognized.

—GEORGE ORWELL, *THE ROAD TO WIGAN PIER*

TEN MINUTES INTO my visit to the Food Innovation Laboratory at the U.S. Army Research Development and Engineering Center in Natick, Massachusetts, a robot named Foodini starts to disobey military orders. Michael Okamoto, a mechanical engineer, and Mary Scerra, a food chemist, are doing a test run of Foodini for their boss, Lauren Oleksyk, the director of the lab. The four of us are huddled around the bot, which is the shape and size of a large microwave and has been programmed to create edible objects using 3-D printing tools. We're peering inside Foodini's glass window, riveted by the seemingly simple task it's about to perform: the production of a two-layer snack that, if all goes smoothly, will come out looking like an open-faced sandwich topped with an army star, printed in avocado.

Most 3-D printers (also called "maker bots" in geek circles) build objects with plastics. They deposit liquid polymers in dots or lines that quickly solidify, adding layer upon layer to form virtually any shape imaginable, from rubber duckies to intricate machine parts to DIY firearms. In 2016, Oleksyk proposed that Okamoto convert a commercial 3-D printer into a model that they could build with edible pastes. He did so, and then worked with Scerra to print a variety of geometric desserts: chocolates the shape of honeycombs and hexagons, spiral nuggets of marzipan, and a deconstructed Reese's Peanut Butter Cup made of chocolate layers topped with a cube of nutrient-fortified peanut butter. The outcome of their experiments are beautiful—and they leave you wondering why this kind of haute experimentation is occurring in the laboratories of the U.S. military.

3-D printed food pellets

"The soldiers of the future will not be eating three-star Michelin desserts in fields of combat," says Oleksyk. "Nor will they be eating plastics, which is usually the first question I get." Her team has been printing the fancy-looking sweets "only because sugary substances are easy to build with." They tend to have optimal "rheology," which refers to the way a liquid flows and its potential to set or solidify. Oleksyk's team is producing these confections as preliminary research for a far bigger and more complex goal, which is, in her words, "to print complete, on-demand meals that can rapidly fulfill a war fighter's total nutrient requirements."

Oleksyk sees these printed meals in much the same way that entrepreneur Rob Rhinehart sees his meal-replacement drink, Soylent, a product I describe earlier in these pages as "adult baby formula." He describes it as an "omnifood," a condensed, all-encompassing form of nourishment that he believes could, in just a few daily doses, sustain humans more effectively than food itself.

Oleksyk knows that if she is to succeed in her mission, though, she needs to be able to print foods more diverse and substantial than chocolate and marzipan. Okamoto and Scerra have been experimenting with ingredients as varied as nut butters, doughs, spreadable cheeses, and vegetable pastes—"substances that are healthy, satisfying, can be flash-cooked or cooled inside the printing chamber, and fortified with extra nutrients," says Scerra.

Foodini is a new machine in Oleksyk's lab—produced in Spain, it's one of the first commercially made printers in the world designed for food production. It's still, however, an early-stage product that can be glitchy and unpredictable. Inside Foodini, the printing chamber is spacious and mostly empty, except for a robotic arm, which descends from a pole mounted to the ceiling, and a rack on one wall holding several inch-thick cartridges, also called syringes, that contain soft or liquid foods. Scerra is using only two cartridges for this test run; one is loaded with a dough made with pea-protein flour, the other with a paste made from avocado. At the bottom of the chamber, there's a glass plate onto which the food is printed. The plate can be quickly heated or chilled—one of several mechanisms Foodini uses to cook, cool, dry, or otherwise solidify the ingredients as they're printed.

We watch, enthralled, as the robotic arm selects the cartridge of pea-protein dough that will form the base layer of a mock sandwich. Through a narrow nozzle, the bot begins to deposit a steady stream of dough, tracing the perimeter of a square the size of a bread slice. It then begins to fill in the square with contiguous lines of dough until, abruptly, the dough stops flowing—even as the arm continues to move.

"Looks like we've got a clog," says Scerra.

"Could be a hole, an air pocket," says Okamoto. The syringe sputters and the arm unexpectedly retracts to a corner inside the chamber.

Foodini then has what appears to be a mild robotic tantrum, squirting a messy pile of dough at the edge of the glass plate.

"It's purging the clog!" says Okamoto. Then, instead of going back to fix and fill in the dough layer, as Foodini's been programmed to do, it jettisons the cartridge.

"It didn't obey!" Scerra exclaims. "It's going for the avocado!"

Now wielding the avocado cartridge, Foodini's arm maneuvers into the middle of the chamber and proceeds to deposit the lumpy, green likeness of an army star on top of the malformed flatbread.

Oleksyk, who is fifty-five, with wispy gray-brown hair, bright blue eyes, and a gentle demeanor, is unfazed by the results, just as Jorge Heraud was upbeat when watching See & Spray make mistakes during its early trips through the Arkansas cotton fields. Foodini, like See & Spray, is the machine equivalent of a toddler. "These 3-D printing tools will evolve very quickly," Oleksyk assures me. She's confident that her grand plan to print nutritionally complete meals for soldiers is not far out of reach. "We can achieve it in the next decade, and possibly sooner."

And while she appears to be pursuing a radically different approach to food security than, say, champions of ancient plants like Wes Jackson and Mark Olson, they all share a similar goal: higher-quality nourishment. "What I like about 3-D printing is its potential to bring freshness and nutritional integrity to military rations," Oleksyk says. "We can produce purer, cleaner, customized, better-tasting foods that are orders of magnitude healthier than what combatants get now, with much less waste."

LAUREN OLEKSYK GREW up the fifth of six children in the small town of Uxbridge, Massachusetts, where her father was a foreman at the local power company and her mom ran the household. On Christmas, each kid in the family got one gift, and in December 1970, when she was seven, Oleksyk tore open a cardboard box that contained a make-your-own Tootsie Roll kit. "Toot Sweet" had a mixer for the chocolate paste, a press to make the rolls, and casings for packaging. "I was completely mesmerized," she remembers. "That's what started me down this path." Every Christmas thereafter Lauren got a different food-processing toy: a popcorn machine that cooked the kernels and melted butter with the heat of a lightbulb; a kit for making gummy dolls with special molds for making gummy dresses. "I didn't eat them," says Oleksyk. "It wasn't the candy that I loved so much as the process behind it." Naturally, she read Roald Dahl's *Charlie and the Chocolate Factory* "about a hundred times—Willy Wonka was like my Thomas Edison." To this day, she refers to her lab as a "factory filled with Wonka wonders."

Oleksyk's family also approached DIY food production in a more practical way: they ate most of their meals from a huge, two-acre backyard vegetable garden stocked with sweet corn, green beans, snow peas, potatoes, onions, tomatoes, peppers, apples, rhubarb, berries, and herbs. They froze and canned enough of it to feed them through the winter months. In summer, Oleksyk's mom would have her run out to get corn for dinner literally seconds before she cooked it—when the water started boiling. "Ripe corn starts developing starches and losing sweetness in its kernels as soon as you pick it," Oleksyk reasons. She learned to assess ripeness and flavor quality for everything they grew, and understood from both her Christmas gifts and her garden-picking that, whether fresh or processed, "food is chemistry."

After high school, Oleksyk enrolled in Framingham State College near her home, which has a strong program in food science. Throughout college she worked in the chemistry department and at nineteen she got a summer job at the U.S. Army R&D Center in nearby Natick. "I got lucky—it was the early eighties, a boom time in food science." The Natick center, which has about 1600 employees, is responsible for producing all the clothing, gear, and shelter used by soldiers in the field, in addition to their food. When Oleksyk arrived, the scientists employed in the Combat Feeding Directorate, which is tasked with feeding all branches of the military, were trying to find a way to make rib-sticking entrees like "meatloaf with gravy" and "turkey tetrazzini" shelf-stable without refrigeration for three years. The results of this effort were MREs (Meals, Ready to Eat). Packaged in thick plastic bags, MREs are hardly appetizing—precooked, ultrapasteurized meals made with unrecognizable meats suspended in treacly sauces and pumped with chemical preservatives. But they were a step up from the rations of previous decades—heavy cans of Spam-style meats and pouches of freeze-dried foods that couldn't sustain rough handling (they're easily crushed). Oleksyk quickly found a mentor in Dan Berkowitz, a chemist developing a bread that could accompany the MREs and remain "fresh"—soft and mold-free—over time. In a matter of months, Oleksyk had helped him crack the code on a "three-year bread" recipe, for which she then received a patent and a full-time job offer.

Thirty-two years later, Lauren is still working in the same laboratory, running a team of chemists, biologists, and engineers who are facing a whole new set of challenges. In the interim, she's helped achieve many firsts for the Combat Feeding Directorate—the first flameless ration heaters for warming up MREs in the field (another of Oleksyk's patents), the first oxygen-scavenging plastics—a polymer packaging material that sucks away oxygen to keep food fresh longer, the first tube foods for high-altitude pilots, and the first high-melt-point chocolates, which stay solid in hot environments. More recently, her team produced the world's first "three-year pizza," as she calls it, which adds sauce, cheese, and pepperoni to the successful bread concept she developed in the early 1980s.

It's an incredible feat of chemistry to keep vegetables, meats, and cheese from getting soggy and moldy without refrigeration for years on end, Oleksyk tells me. "You ask any soldier how we can improve rations and they'll say 'the thing we miss most out there is pizza.'" But even this doesn't compare in terms of novelty or potential benefit to what lies ahead. "We're entering the most extraordinary moment of innovation in my three-decade career—the last couple of years, especially," she says. "With the advent of robotics, sensors, and big data, things have just exploded."

Oleksyk first became interested in 3-D printing when exploring methods for customizing food for soldiers with specific nutritional needs. Her colleagues at Natick were beginning to integrate Fitbit-style sensors into the fabrics of soldiers' uniforms that could generate real-time data about a soldier's health status. The air force was developing sensors that could be taped to the skin of pilots to regularly analyze the biochemistry of their sweat and detect fatigue. "We're moving into an era of increasingly detailed biometric data on each soldier," says Oleksyk. "Using sensors and genomic information, we'll be able to monitor their exhaustion and stress levels, we'll know more about their microbiome health, their immune strength—all this will inform our understanding of their nutrient needs." The personal data, she posited, could be transmitted to a 3-D printer, and the printer in turn could play the role of a futuristic apothecary, blending customized supplements into the food pastes as it prints bars and pellets for each soldier. Maybe Soldier X is low on potassium and needs a calorie boost, so the printer incorporates oils and powdered sweet potato into his dough, while Soldier Y needs extra calcium and vitamin C—in go her supplements. As Oleksyk envisions it, the personalized food items will then be delivered by drones from the nearest base camp to soldiers deployed even in the

Our unsuccessful attempt to 3-D print hummus flatbread with an avocado star

most remote regions of combat. So while the ingredients haven't exactly been freshly picked, the nourishment has at least been freshly printed.

The concept of personalized nutrition did not originate in Oleksyk's lab. Practitioners of Ayurvedic medicine have been adjusting diets for individual *doshas*, as my yogi friend informed me, for thousands of years. More recently, food giants ranging from Nestlé to Campbell's Soup Company have invested in start-ups like Habit and Freshly, which are formulating personalized diets for their customers based on genomic testing. "The most significant trend of the future will be personalized nutrition," Denise Morrison, the former CEO of Campbell's, tells me. "It will become much more affordable and accessible to monitor your individual nutritional needs and eat a customized diet."

Oleksyk, for her part, is collaborating with private-sector partners—including the Dutch research organization TNO, a world leader in 3-D printing technology—to accelerate her progress. "Everything we do is with an academic or industrial partner," she tells me. "The goal is to make this research beneficial first to soldiers, but ultimately to every household in America."

IT'S NOT JUST rapid changes in technology that have propelled Oleksyk's latest R&D projects, but also physical and demographic changes within and among the soldiers themselves. Oleksyk's rations feed about 2.1 million military personnel per year. The average age at enlistment is just under twenty-one, the maximum age allowed is thirty-nine—this is Generation Z.

"We've seen a larger shift in the dietary needs and desires in this generation of soldiers than in any other in my three decades here," Oleksyk tells me. Some of it is demographic and cultural—there are now more Latino enlistees than ever before, more Muslims with halal diets, more vegetarians, more soldiers concerned about food and packaging waste (a lot of both gets wasted when it comes to MREs), and more soldiers concerned about product labeling, healthy ingredients, nutritional density, and high-performance diets.

"The soldiers we're now feeding are the Fitbit generation—they're comfortable with monitoring their vitals, their health status. They're also much more interested in performance foods—they come in saying, 'We want to eat what will make us better fighters,'" Oleksyk says. "They don't want chemicals in their food, not a lot of sugar. The requests we're getting are for things like 'real fruit juice' and 'no mystery meat.' You didn't hear that even five years ago."

This is also, though, the first generation raised on smartphones and tablets, which has had adverse impacts on their physical resilience. "It's a striking contradiction—the soldiers we're feeding today are more concerned about what they eat but they're also, at a baseline, less physically fit," says Oleksyk. "They grew up more sedentary than previous generations. They didn't walk and bike around the neighborhood like you and me, they spent a lot more time sitting on the couch looking at devices. They grew up taking more antibiotics and eating more processed foods than we did—more mac and cheese." All these lifestyle shifts have caused measurable changes in the soldiers' physiognomy. Oleksyk has heard from researchers at Fort Bragg that the current population of enlisted military has noticeably lower bone density than previous generations—a result of both less consistent physical activity and less nutritious diets. "What we know for a fact is that when soldiers go to basic training they end up with more injuries; we're seeing stress fractures at a much higher level than soldiers in the past. And it's up to us to augment their diet, to increase their bone health," she says.

There's also a lot Oleksyk doesn't yet know. Military scientists are testing fecal samples to understand soldiers' microbiome activity and the impacts of antibiotics on their gut health. The researchers are testing how performance nutrition (nutrient-fortified bars) affects soldiers' cognitive abilities during high-intensity physical events—for example, target accuracy during live-fire practice at the shooting range.

Oleksyk points out that soldiers are a representative sampling of their entire generation, and Gen Z now accounts for about a quarter of the American population—a larger cohort than the surviving Baby Boomers and Generation X. "These cultural and biophysical trends are true for many young Americans, not just the enlisted," she says. "I think it's a safe bet that future nutrition will become increasingly customized for soldiers and civilians both."

But Oleksyk has still another factor to contend with: military strategy is changing. Soldiers today are deployed on decidedly different types of missions than they were in World War II, Vietnam, or even Iraq. Historically, soldiers ate en masse at base camp or even near the front lines. During the Civil War, they actually had bread bakers on the front lines (Oleksyk shows me an old photograph of young women wearing masks while kneading bread near a dugout). By World War II, they'd moved to canned meats but they also set up mobile cafeterias that fed hot meals to hundreds of soldiers at a time. Even during the Gulf War and the Iraq War they established pop-up kitchens for feeding soldiers in large groups. Now increasingly the military strategy is moving to smaller, more mobile, and decentralized squads. "Soldiers will be on missions where they can't carry a lot of weight, where they'll be moving in smaller groups and may be gone from base camp for much longer periods of time," says Oleksyk.

They will also, more than likely, be in environmentally stressed locations. Military hot spots are under increasing resource pressure. "We can't expect that they'll be able to find fresh food or forage for edible plants," she reasons. "They'll be in situations where everything that sustains them will have to be shipped in and out, including water." If you consider these factors together, Oleksyk's adventures with Foodini begin to make sense.

IN THEORY, A PRODUCT like Soylent, which has been touted by American football players and CEOs alike as an effective meal-replacement product, could meet the dietary demands of next-gen soldiers. Oleksyk routinely explores the private sector to see whether there are commercially produced options that can serve her goals. But Soylent in the liquid form is too heavy for soldiers to carry on extended missions, and the powder form can be dehydrating—a challenge if you have a limited water supply. The same goes for the freeze-dried Chicken-Flavored Pot Pie I sampled after visiting the Wise Company headquarters in Salt Lake City: if you're a soldier in remote Afghanistan, a freeze-dried entree takes up too much space in your backpack and requires precious water to rehydrate, and it can easily turn to powder. Oleksyk adds that she can't sustain soldiers on liquids and mush alone: "They need to chew." A product Oleksyk likes is KIND bars—they're compact and nutrient dense, but they're also sugary, likely to melt, and well outside of her price point.

Ann Barrett, a biochemist on Oleksyk's team, says the main purpose of her research is finding ways to pack the highest amount of calories and nutrients into the least amount of edible space—so as to minimize what soldiers have to carry. Barrett has invented a hypercompressed activity bar that she calls the Sonic, which NASA recently adopted for its Mission to Mars menu. The Sonic has more than double the calorie density you'd get in a standard performance bar, and it doesn't rely on the sugary syrups or chemical binders to cohere the ingredients. Barrett used a binding technique called sonic agglomeration, which she says "densifies" ingredients without affecting their nutrient and flavor profiles.

The bars come in flavors that range from coconut-almond to jalapeño cheddar. Barrett explains the process in zealous geek-speak: "We bombard the ingredients with ultrasonic waves, which resonate at a higher frequency than the upper audible limit of human hearing. This causes the food particles to vibrate and shift, increasing the surface area for interparticle connection." Translation: the process is a bit like shaking a bucket of ice so the cubes unstick and settle together. Once the food particles are optimally aligned (which takes milliseconds), the mixture is then subjected to enormous physical pressure as it's stamped into bars.

This sounds to me like a lot of complexity for a few square inches of food, but when I bite into the coconut-almond bar, I'm won over: the texture is firm and somehow velvety, far more palatable than the gummy consistency of a PowerBar, and it's rich and nutty-tasting, salty-sweet, but not too much of either. Added bonus, says Barrett: "If you have hot water, you can dissolve the bar to make coconut soup."

Another food chemist on Oleksyk's team, Lauren O'Conner, has developed a novel way to fulfill soldiers' request for real fruit juice using a process known as radiant zone drying, which she describes as "kind of a gentle form of freeze-drying." Typically, a soldier's field rations contain powders for making Kool-Aid and Tang—lots of sugar, zero nutrient value. Soldiers crave fruit because it's loaded with phytonutrients; it's also highly perishable and without refrigeration can't last in the field. Phytonutrients are very heat-sensitive and easily erode when food is dried or pasteurized. "Radiant zone drying preserves those really sensitive nutrients—we purée the fruit, and expose the purée to extremely low heat and pressure to cause evaporation," says O'Conner.

The result looks like a fine, fiesta-colored confetti in apple beet (a deep purple), tropical citrus (a sunny orange), and strawberry banana (a peachy pink). "The colors are preserved because there's been no damage to the nutrients," says O'Conner. She mixes the confetti with some cold water and, again, I'm impressed. The stuff tastes as advertised—bright and fresh. And with Oleksyk's special oxygen-trapping pouches, it can stay that way for three years.

Dr. Tom Yang, a senior member of Oleksyk's team, has been experimenting with a method that may be more promising than any other. Microwave vacuum drying carefully controls the dehydration process, transforming food not into the Styrofoam consistency that results from freeze-drying, but into a firm texture that lies somewhere between dried apricot and solid Parmesan cheese. "You remove moisture only to the point where the food is completely stabilized," Yang explains. "No bacteria can grow, and the flavor and nutrients are almost completely retained." This method, like a standard microwave, dries food from the inside out (rather than like an oven, which heats or dries from the outside in) and allows for careful control of the process. You can use this approach to partially dehydrate almost anything—a chopped salad, a wedge of Brie, an omelet, a plate of French toast, a bean taco, a bowl of mac and cheese. The texture may be off-putting to some, but when Yang offers me an omelet that's roughly the consistency of licorice, I find it oddly delicious and fun to chew. After the microwave-vacuum-drying process takes place, the food is about a third of its original size. Foods with high water content, like fruits, can be reduced down even further.

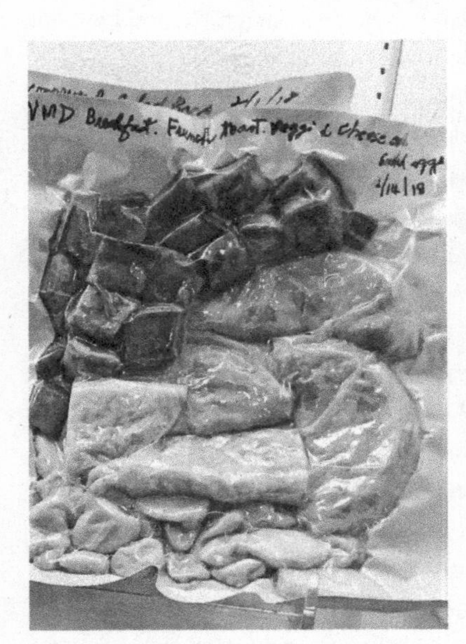

French toast, omelet, and scrambled eggs—vacuum dried

At the end of my visit, Yang gives me a goodie bag of his creations. "This has fifteen fresh strawberries and about twenty blueberries—nothing else," he says, pointing to a bar just bigger than my index finger wrapped in plastic. "And here's a large salad," he says, handing me shrink-wrapped greens about the weight and thickness of a chequebook.

On the way to Logan Airport from Natick, I eat one of Dr. Yang's vacuum-dried orange bars. My plan had been to stop and get a steaming bowl of pho in East Boston before my flight, but I get mired in traffic. Bumper to bumper for miles. I haven't slept or eaten much, and am experiencing an extreme sugar low when I remember that Yang had mentioned the bar "contains six whole oranges" and is supercharged with natural vitamin C. I tear the plastic casing with my teeth. The stuff inside tastes extremely tart, explosively orangey, and is roughly the texture of a melon rind. The effect, though, is nearly miraculous—

something akin to the scene in *Pulp Fiction* where John Travolta stabs Uma Thurman with a syringe full of adrenaline and wakes her from a coma. I feel like I've had a total-system reboot—and like I might actually be warming up to this concept of meal replacements.

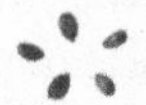

SOON AFTER I get home from the Natick lab, I head to Amazon.com and order a case of Soylent. I'd tried to buy it locally, but it's sold out at Target and Walmart—chains that now sell it nationwide. My Kroger tells me "we carry it, but not in Tennessee." Whole Foods is hostile: "We don't do Soylent," a store manager sniffs. I see why when my case arrives: The outside of the cardboard box that holds my twelve-pack proclaims in bold lettering: "Proudly made with genetic engineering."

Rhinehart, Soylent Inc.'s founder, has deliberately positioned the product as a kind of antihero of the sustainable food movement. The omnifood is glibly named Soylent after the 1970s sci-fi movie *Soylent Green*, in which the characters survive in a post-food dystopia on a brew made of blended humans. A thirty-something denizen of Silicon Valley, Rhinehart founded the company in 2013 with three other technoids whom he met while working on start-up projects in Sunnyvale. Rhinehart and his associates had been working eighty-hour weeks and subsisting on the classic frat-boy diet of frozen corn dogs and instant noodles. They began to try to "hack a recipe" for an instant, chuggable nutrition source that would save them time and obviate the need to eat. Rhinehart open-sourced their formula and over many months, with the help of other nutrition hackers around the world, developed a blend of fats, fibers, carbohydrates, and thirty-five vitamins and minerals. He's since raised more than $70 million to fund the start-up. "Food, reformulated" is the company's tagline. Each 8-ounce bottle costs $3.25, has 22 grams of fat to keep hunger at bay, and supplies 20 percent of your total recommended daily nutrition (per the U.S. Food and Drug Administration, the Department of Agriculture, and the National Academy of Medicine). If you consume five bottles a day, you get everything your body needs to "thrive" and "a steady hum" of energy all day long, according to some of the thousands of reviews on Amazon.

Rhinehart describes Soylent with a kind of messianic fervor as "a way to disrupt our entire food system: Food is the fossil fuel of human energy. It is an enormous market full of waste, regulation, and biased allocation with serious geopolitical implications," he wrote in 2013. He envisions a world in which his drinkable formula is a "civil resource" that gets piped into your house through taps like water, and where, due to climate-change pressures and global conflict, eating real food will become recreational. He says, "We'll see a separation between our meals for utility and function, and our meals for experience and socialization."

Both the product itself and the way Rhinehart frames it seem to lack humanity—and, for that matter, joy. Many of Soylent's critics have noted this, with varying degrees of humor. When Rhinehart appeared on *The Colbert Report* in 2014, soon after his company launched, comedian Stephen Colbert asked, "What was the inspiration for this? Did you see someone in a coma with a feeding tube, and you thought, I'll have what he's having?" A *New Yorker* article on Soylent titled "The End of Food" quotes Walter Willett, the chair of the nutrition department at the Harvard School of Public Health, saying that the formula ignores, among other things, the value of phytochemicals present in fresh fruits and vegetables such as lycopene (which makes tomatoes red) and flavonoids (which make blueberries blue). "It's a little bit presumptuous to think that we actually know everything that goes into an optimally healthy diet," Willett said. He added that surviving without plant chemicals is biologically possible, but "you may not live maximally, and you may not have optimal function. We're concerned about much more than just surviving."

Maybe it's because soldiers in fields of combat receiving military rations *are* concerned, principally, with surviving that I found Oleksyk's food-replacement technologies intriguing more than frightening. And yet my first reaction to Soylent—same concept, for civilians, not soldiers—was, as it may be for most of us, skepticism with an undercurrent of nausea. Soylent seems to represent, if not the death of food, then the death of the *soul* of food. It's pabulum, literally and otherwise: bland, simplistic, insipid. I grew up in a household where food was a proxy for love. My mom, a psychoanalyst, meted out affirmation throughout my childhood with restraint. She was not at all restrained, however, with the flavors and aromas in her kitchen. To this day, she loves to cook, cooks with love, and to eat her food is to feel lavishly adored. For these and other reasons, a foodless world seems to me, as it does to many of us, like a loveless world.

And yet more often than I'd like, if inadvertently, I eat soldier-style—simply to quell hunger and stay on mission. I have, if I'm honest, many moments like the one stuck in Boston traffic on the way to the airport, when I'm unable to sit down to a proper meal and then resort to whatever packaged food product is near at hand. Breakfast for me often consists of whatever remnants are left on my kids' plates, eaten standing while washing dishes and checking email. Lunch is often a bowl of cereal, spoonfuls of peanut butter, leftovers that the rest of my family won't touch, or some forgettable take-out item from a plastic clamshell. It's fast, convenient sustenance that's usually bland and almost never close to nutritionally complete.

My mother and grandmother, who put so much loving attention into the food I was raised on, typically spent several hours a day on the shopping, preparation, serving, and cleaning up of our family's daily meals. If I'm honest, my husband and I allocate about a fraction of that time to preparing food for ourselves and our kids. Family dinner is the one meal we still battle to protect. A few times a week we set aside a full hour or two to cook and eat a meal together and let food do what it does so well—engage the senses, ground us in our home, connect us to one another.

So despite my aversions to Rhinehart's adult baby formula and his marketing ploys, the reality is that I'm already living the life that he describes, in which there's a clear separation between the fast, bland meals eaten for "utility and function" and those eaten with family and friends for "experience and socialization." That's why I placed the Amazon order: I'd begun to see that Soylent might be a reasonable stand-in for—even a better alternative to—those half-assed meals I regularly consume on the "utility" side of the equation. It seemed to me that Soylent—and, eventually, a Foodini in my kitchen—could raise the nutritional quality of my grab-and-go breakfasts and lunches, maybe even liberate time and energy so that I could invest more of both in the one meal a day I sit down and eat with my family for pleasure.

There's also the fact that Soylent is vegan, extremely low-carbon, cheaper than a McDonald's combo meal, and results in zero food waste. By purely ecological and socioeconomic (if not culinary) standards, this soulless stuff may be a truer embodiment of a "sustainable, equitable food" product than a locally grown six-dollar-per-pound heirloom tomato.

THE FIRST BOTTLE of Soylent I drink takes some effort. It tastes exactly as the reviews on Amazon said it would, like a cross between almond milk and pancake batter, neither of which inspires in me the desire to chug. By the second bottle I get used to it—a far cry from delicious, but oddly sating. By the third, I start looking forward to the next one. I welcome it with the kind of relief I feel when I push the Off button on my cell phone. It silences, if only for a meal a day, or a few meals a week, that noisy inner monologue many of us have about what we should eat, whether it will satisfy our cravings, whether it's affordable, responsible, and so on. We live in an era where many if not most of us overthink food. We fetishize it or we fret about it or both. It's a relief to experience some freedom from all that, and while Soylent is not a solution, it's at least an opportunity to press Pause.

"A nutritionally-complete staple food designed to provide maximum nutrition with minimal effort"—that's the product description on the Soylent website. It echoes, uncannily, the comment Columbia University professor Ruth DeFries made at the beginning of this journey: "Every new agricultural tool introduced since the first farming settlements has been designed with the same goal: to coax more food from the earth with less human effort."

Whether we like it or not, Soylent is another link in that long chain of technological experiments stretching over ten millennia that's aimed at stretching our food supply, producing more staple nutrition with less effort.

This brings me back to that question I was searching for in the vat of freeze-dried vegetable chunks at the beginning of the book—and the same question that came to me as I huddled with Lauren Oleksyk and her team around Foodini, peering inside its glass window as the baby bot bungled a futuristic snack: How screwed *are* we? Where is this long chain of technological experiments going, exactly? Can we reasonably, responsibly dare to hope, given the failures of technology in the past and the environmental and population pressures to come, that it's headed in a good direction? We want assurance not just that there will be enough food for all of us to survive, but that our culinary traditions, including our fresh-food supply, will continue to live on.

The answer, in so many words, is yes. It's far more likely than not that there will be enough food for all of us, and that we'll protect and preserve our food traditions. The answer is that we're no more screwed than we are capable of rising above the challenges we face. This long, strange trip I've taken through eleven countries and thirteen states was finally a search for hope, and at the end of the day I found it.

For at least the foreseeable future, many of us will have choices, even more than we do today, about the kinds of food we want to eat, about what aspects of an evolving food system we want to support. Chances are good that if you want to continue eating local, organic, and artisanal foods nourished by soil and sun, and you're willing to pay a premium for them, you can go on doing that for a long time to come. And if you want to live on a diet of hyper-personalized nutrition, eating customized foods for optimal health, that will also be an option. And if you want to stop cooking and chewing food altogether, you can subsist entirely on Soylent (or another drinkable nutrition) for about fifteen dollars a day, automatically shipped to your door.

For me, this book has been as much a process of shedding my own old hang-ups and misconceptions about food as it's been an exploration of new ideas and future possibilities. I've gotten over some of my nostalgic fantasies of being a DIY backyard farmer and of eradicating industrial agribusiness. I've let go of my misconceptions about gene-edited foods, farmed fish, petri-dish meats, and even adult baby formula. I've become hopeful even about the advent of 3-D-printed foods and Soylent. Even our fast foods, the stuff we eat quickly and carelessly, will change for the better, and, in turn, our respect for traditional foods will deepen.

There will be trade-offs. The underpinnings of our food system—the methods and tools and techniques future farmers use to grow fruits, vegetables, grains, and proteins—will change, in some ways subtly and in other ways radically, in order to continue growing the traditional foods we love. We'll need passionate grassroots activists who continue to protect those traditions, and stronger state and federal policies that guide farmers toward smarter, more efficient practices. We'll need robust networks of local, organic, small-scale farms, but also large-scale industrial farming, done better. We'll need smart fish farms and AI-enabled robots and good GMOs and CRISPR'd crops just as much as we'll need to safeguard heirloom plants. We'll need rich, healthy topsoil, but also the data gathered from intelligent sensors planted beneath the surface. We'll need new scrappy little start-ups and old, big food companies pulling and pushing for a third way approach to sustainable food production that serves everyone, not just the wealthy elite. We'll need to push the bounds of technology with a better understanding of where it has failed us. We must innovate—with humility.

Epilogue

Growing Up

IT'S 8:15 ON a Sunday morning in May, chicken-slaughter day, when I arrive at the farm of Chris and Annie Newman, an eight-acre spread beside an old grove of oak trees at the edge of the Potomac River in northwestern Virginia. Annie, who is thirty-two, is tired. She was up most of the night with Betty, their one-year-old, and woke up again before dawn to water and feed the 300 hens, gather their eggs, cook the family breakfast, and ferry their three-year-old to neighbors. Chris, who is thirty-six, got up even earlier to feed the hogs, and cull from their brood the 150 broiler chickens that he and Annie will now kill, process, package, and freeze before sundown.

I meet them in their makeshift slaughterhouse—a 600-square-foot metal trailer that's parked in the driveway of their small clapboard home. The trailer has a *Breaking Bad* vibe with its outdoor generator, stainless-steel tables, baths of cold and steaming water, and obscure-looking machines, one a giant colander with thick rubber fingers (a defeathering device). There's a tall circular stand with eight "killing cones"—stainless-steel vices used for humane poultry slaughter. A stack of blue plastic crates hold the chickens, which are huddling quietly, their red combs and scraggly white feathers covered in dust. Annie grabs each plump three-pounder by its feet, flips it upside down, and inserts it so that its head dangles out the narrow end of the cone. One by one, she grips the throat and slices the jugular.

"Shit."

A runty bird has wriggled through the end of a cone and dropped into the deep puddle of blood that's draining into a plastic basin below. Chris looks at the drenched creature sputtering in the basin. "Oh, babe, you got a little tiny one," he says.

"I think he's sick—something's wrong with his leg," Annie responds.

"Want to mercy-kill him?"

"Yeah. I just, I hate it when this happens."

Annie kneels down and decapitates the bird. Dark-haired and pale-skinned with delicate features, she looks like Katniss Everdeen crossed with a Jackson Pollock painting. From her purple headscarf down to her battered combat boots, she's splattered with fresh red chicken blood. "I'm used to getting dirty," she says. Annie is a painter who grew up visiting building sites with her dad, a contractor. "I just don't like to see a life end like that. No matter how many kills you do, that kinda thing will gnaw at you."

And so it should, says Chris: "We never want to get to the point where we stop feeling that. If we do, we should pack it in and go back to our office jobs."

Chris, who is six foot four with a shaved head and sinewy arms, wears a shirt under his butcher's apron that reads "Meant for More."

His dad grew up just across the Potomac, in the Chesapeake region of southern Maryland, which is still home to thousands of Piscataway, the Native American tribe from which Chris and his dad are descended. His mom, who's African American, raised Chris in southeast Washington, D.C. As a young kid he was a fast learner who by the second grade was reading at a twelfth-grade level. "It's not lost on me that I've taken an unusual path in life," Chris tells me while gutting a bird. "I was that rare, nerdy, biracial kid raised in a black neighborhood a long-ass way from a Whole Foods who winds up at the age of thirty-six with his hand up the butt of an organic chicken."

Of all the people I've met throughout this story, Chris and Annie will strike me, during the course of my visit and our many conversations thereafter, as the most heroic. Not because they're the most likely to succeed—to feed the greatest number of people or make the biggest impact or win the Nobel Prize—but because they represent the most interesting and authentic synthesis of ideas about third way agriculture I've encountered.

Together they share a kind of revolutionary fervor, a determination as small farmers to challenge and reimagine the status quo of sustainable food production. Day after day, they test out their ideas in the trenches as they try to build a new paradigm for small-scale food production that synthesizes tradition and technology, the very old and the radically new. The more I get to know them, the more I begin to see their vision for what I can only describe as their own personal Wakanda—a food-rich forest ecosystem that will eventually be managed and tended by intelligent machines, a place where technology exists to serve and elevate nature.

OCCASIONALLY THROUGHOUT HIS childhood, Chris and his dad visited the house Chris now shares with Annie and their girls. Chris's dad volunteered with the local Indian Tourism Association, teaching classes on the history and culture of indigenous groups, about how they lived on the land, how they ate, their values and customs. Chris's paternal grandmother cooked with ingredients native to the region—pawpaw, persimmon, chestnut—and taught him about forest-based agriculture. "For centuries, the Piscataway managed dense food forests that produced fruits and nuts in the upper levels and understories of trees, in the bushes, on the vines and in the ground cover of these forests—you couldn't take a step without tripping over a meal," says Chris. "I was taught early on that we live within the ecosystem, not on top of it. That we exhale, and the trees breathe that in. That our ancestors die, we bury them in the soil, their bodies over hundreds of years nourish the system. Plants and animals decay and are reborn. Everything living comes from something that died and regenerates."

None of this was top of mind for Chris during his twenties, when he graduated from the University of Maryland with a degree in computer programming and got a lucrative job at Lockheed Martin. He later joined a team building software for the Treasury Department. "The work was not exactly edifying," he deadpans. When he was twenty-eight, Chris met Annie at a dive bar in the Adams Morgan neighborhood of D.C., where she was working at an art gallery. "He walked over, *looked* at me with his green eyes, and said, 'You're smiling like you've just found cake in your pocket,'" Annie recalls. "And that was that."

Emboldened, Chris moved to a smaller tech firm, but the hours were grueling and he started to experience stomach pain. During a vacation, Chris read a neighbor's copy of Michael Pollan's *The Omnivore's Dilemma,* where he was introduced to the concept of permaculture. He felt "a sudden and overwhelming sense of recognition, like if you've known something your whole life and then somebody comes along and gives it a name," he says. Permaculture is based on the principles of indigenous food production Chris had been taught as a kid. The concept also resonated deeply for Annie, an Episcopalian. "The Bible talks a lot about stewardship of the land," she tells me. "Chris and I were coming at this from two different belief systems, but kind of meeting in the middle."

Pollan had profiled a permaculture pioneer named Joel Salatin, who managed a farm a few hours away, in the western part of Virginia. Days after reading Pollan's book, in the summer of 2013, Chris quit his job, enrolled in one of Salatin's workshops, and decided with Annie to start a farm they would later name Sylvanaqua, meaning "the forest next to the water."

The farm got off to a rocky start. "Of course, we had no frigging clue how hard it would be," says Chris. The only person in either of their families who'd known anything about farming was Chris's late maternal grandfather. "I had agriculture in my ancestral blood, from slaves and tobacco farmers and native forest dwellers—problem was, none of them were around to tell us what to do," Chris tells me.

They learned what they could from a few books and a lot of YouTube videos. But they made some serious miscalculations, most notably about their path to profitability. They thought they could produce a decent income within six months; instead they lost money for four years. When I visited them in the summer of 2018, they were just breaking a profit. They'd tripled their production over 2017, and in an average week they were harvesting about 1800 eggs, 180 chickens, 1200 pounds of pork, and 600 pounds of beef—all of it organic and sold in high-end, D.C.-area stores and restaurants. It would yield them about $170,000 for the year, but their net was about $28,000—a poverty-line wage for parents of two young children.

They'd gotten used to padding their income with side jobs. A year after starting the farm, Chris got a part-time gig working remotely for that software company he'd quit—a job he still does about twenty hours a week "to keep the lights on." Annie took a job clerking at Belk, a discount department store, but left when they started having kids. Parenting pressures inevitably added to their professional pressures. Soon after their younger daughter was born, their three-year-old, Marlenne, was diagnosed with autism; managing her therapy complicated their already complex daily regimen.

"The romantic-sounding stuff about farm life is definitely very romantic—working together in the fields, watching your kids out chasing chickens," says Annie. It's less fun, though, when a coyote massacres half of your hen flock, or a torrential storm comes and you've still got to do the outdoor chores, or it's 105 degrees [Fahrenheit] outside and 99 percent humidity and it hasn't rained in a month.

Chris and Annie occasionally talk about quitting—"sometimes more than occasionally," he admits—"but then I get that knock on my conscience, like: you don't get to quit this, punk. We're here to help solve a systemic problem." Their mission at Sylvanaqua, as Chris describes it, is twofold: to prove that the best way to produce food is "by mimicking timeless patterns in nature"; and to use whatever advanced high-tech tools and techniques can help them do that.

A FEW MONTHS before my visit, in 2018, Chris had begun to question the efficacy, and even the ethics, of the sustainable food movement. It had begun to bother him that nobody in his community, including most of his friends and the neighbors he grew up with, could afford the $10-a-pound pork chops and $4-a-pound chicken he and Annie were selling to high-end markets and restaurants—food that realistically can feed only the rich.

Chris had found himself splitting away from the philosophies of his mentor Joel Salatin on the grounds that it was elitist and promoted what Chris calls an accessibility gap. "A common refrain among these folks is that poor people should value food more, should be willing to pay more," Chris tells me. "Well, I grew up with poor people and I'm not going to go 'That kid in the projects, he needs to value food more.' What we need to figure out is how to grow sustainable food in a better, more equitable way, and that means bringing in smart tech."

As a guy trained to code, it's not surprising that Chris is both pro-ecology and pro-technology. "There are lots of people who call themselves sustainable farmers who see technology as a terrible threat, as the fascist dictator of the natural world and of the modern food system, in particular," he says. "But the problem has never been the technology;

it's been the ethics and the values and the motivations behind the use of that technology, behind why we grow, what we grow, and where we grow. We've been too focused on growing things for a global economy at the expense of local ecologies, and the reality is that if we use it right, technology can help reverse that."

As a software engineer, Chris sees "food as an interface," he tells me. "How it's grown—the methods will become more and more numerous and replaceable. In the future, the same piece of fruit might come from my food forest, or a field in California, or an inner-city aeroponic warehouse. And so long as those growing methods are helping directly or indirectly to restore or relieve pressure on the earth's ecosystems, I support them." Like Ruth Oniang'o in Kenya and Marc Olson in Mexico, Chris and Annie are committed to protecting agricultural traditions using modern methods that sound anathema to many agricultural traditionalists. The Newmans, for example, want to use the robotics and software to help re-create and manage the "food forests" that Chris's grandmother taught him about. They will thin the forests that surround their croplands and grazing areas to create space for multiple stories of apple and cherry persimmon trees. Lower down, they'll plant several stories of berry bushes—huckleberry, blueberry, blackberry, hackberry, and gooseberry. At the ground level, they'll raise mushrooms and bring in their pigs to fertilize the soil. In fields outside the forest, which they will fertilize with the nitrogen-rich manure of their livestock, they'll plant perennial oats, soy, spelt, and wheat; they will intercrop corn, beans, and squashes ringed with sunflowers and herbs.

They'll revive the native plants of the region, using heirloom seeds for the grains and fruits and vegetables they plant—but with some genetic twists. "We're in a completely different climatic situation than my ancestors were in," Chris tells me. "Who knows whatever the heck the climate is going to be in the next ten or twenty years, so genetic engineering will have a place in our plan—in developing local and ancestral seed stocks with genes for resilience to heat and drought and tolerance to soil salinity." Chris and Annie foresee using driverless car networks and eventually drones to transport their meats, eggs, and vegetables directly to customers—a major improvement over their current delivery method, which is a gas-hungry flatbed truck.

Chris and Annie also anticipate more intelligent mechanization on their farm. When they can afford it, they'll invest in robotic weeders and harvesters for fruit and nut trees. Chris also sees "robots becoming very valuable to help us monitor rainfall, sunlight, seasonal shifts, the prevalence of different pests and pollinators, to help us measure whether and when and why certain fruits and vegetables are coming to maturity"—as a means to deepen their understanding of ecosystem management.

I find the vision of this private Wakanda so appealing that I invite my family over to the Newmans' for a future feast. This is where I want to spend Thanksgiving 2050. Chris kindly obliges. The meal, he says, will honor the full spectrum of his and Annie's family traditions—turkey and duck, heirloom varieties of corn, green beans, and potatoes grown at the margins of the food forest, sauces of cranberry and elderberry, along with the plants of his ancestors—pawpaw, persimmon, and chestnut. And every element will have been made possible, maybe even enhanced, by the judicious use of the best mid-twenty-first-century tools.

FOOD FORESTS AND permaculture, Chris cautions, will not be a comprehensive solution for future food production. "It won't solve the entire problem" of feeding everyone affordably, he says. "It will not get us a hundred percent there, not at all." He and Annie envision what they call a "macro-permaculture" system with various food production "zones" that will radiate outward from major urban centers into the surrounding suburban and rural areas. The zone closest to the city, for example, will have a high concentration of aeroponic vertical farms producing the most perishable and high-nutrient produce; suburbs will cultivate large-scale community gardens; exurban areas like their community in northern Virginia will manage the food forests that produce sustainable meats, orchard fruits and nuts, and some grains.

To succeed, such a system would require significant amounts of government funding and an unprecedented approach to, among many other things, land-use and international-trade laws. And even if the vision is realized, says Chris, it will not supplant industrial farms, which will continue to produce, among other things, staple grains, which currently account for more than 90 percent of all agricultural production (some of which, like wheat, may be in increasing demand as we scale up plant-based alternatives to meat). A macro-permaculture system also would not supply the great volume of bananas, avocados, coffee, off-season fruits and vegetables, and other luxuries to which we've grown accustomed that are trucked and shipped over long distances.

What makes the vision realistic, however, is that it does not require a false choice between permaculture and industrialized agriculture, between zero-tech and all-tech. Instead, it supports "a constellation of competing ideas," as Annie puts it. It's also a system that would engage many more participants. In the United States, less than 2 percent of the population currently grows the food produced on our soil. Three decades from now, if we're going to have a healthy, secure food supply, it will require a much larger, more deliberate network of participants, nationally and globally. Whether you're a large- or micro-scale farmer, a gardener, a policy advocate, a permaculture evangelist, a chef, a botanist, an engineer, or a conscientious consumer, many more of us will find a way to participate in the movement to protect and adapt our food supply to the pressures of climate change and growing urban populations.

In doing so, we'll be helping to realize the Piscatawayan motto that's written on a sign hung in Chris and Annie's home:

PEMHAKAMIK MENENACHKHASIK—THE WHOLE WORLD, A GARDEN.

Notes

INTRODUCTION

3 **famine is on the rise:** Rosamund Naylor, "The Elusive Goal of Global Food Security," *Current History* 117 (2018): 3. See also "The State of Food Security and Nutrition in the World," Food and Agriculture Organization of the United Nations (FAO), 2018, https://tinyurl.com/y8jfvy9o.

3 **Nearly 40 percent:** "Health, United States, 2016: With Chartbook on Long-Term Trends in Health," National Center for Health Statistics, 2017.

3 **at least 2 degrees Celsius:** "Climate Change 2014: Synthesis Report," Intergovernmental Panel on Climate Change (IPCC), 2014, https://www.ipcc.ch/report/ar5/syr/.

5 **some 800 million people still suffer:** "The State of Food Security and Nutrition in the World," FAO, 2018.

5 **almost 200 million fewer:** Linda Poon, "There Are 200 Million Fewer Hungry People Than 25 Years Ago," NPR, June 1, 2015.

5 **we spend about 13 percent:** Derek Thompson, "How America Spends Money: 100 Years in the Life of the Family Budget," *The Atlantic*, Apr. 5, 2012.

5 **liberated men and, in particular:** Brent Cunningham, "Pastoral Romance," *Lapham's Quarterly* 4 (2011): 179.

6 **hurting agricultural productivity worldwide:** J. R. Porter et al., *Climate Change 2014: Impacts, Adaptation, and Vulnerability*, IPCC (Cambridge: Cambridge University Press, 2014), 485–533.

6 **2 to 6 percent:** Ibid., 506.

6 **1.5 degrees Celsius:** "Global Warming of 1.5°C," IPCC, 2018, https://tinyurl.com/yb46plrt.

6 **"It's like a deafening":** Quoted in Chris Mooney and Brady Dennis, "The World Has Just Over a Decade to Get Climate Change Under Control, U.N. Scientists Say," *Washington Post,* Oct. 7, 2018.

6 **"Other threats—flooding, storms":** Personal communication, Jerry Hatfield, Sept. 2018.

7 **double by 2050:** Porter et al., "Food Security and Food Production Systems," in *Climate Change 2014,* 512.

7 **more than half of its fruit:** David Karp, "Most of America's Fruit Is Now Imported: Is That a Bad Thing?" *New York Times,* Mar. 13, 2018.

8 **in roughly AD 1200:** Peter W. Stahl, "Structural Density of Domesticated South American Camelid Skeletal Elements and the Archaeological Investigation of Prehistoric Andean Ch'arki," *Journal of Archaeological Science* 26 (1999): 1347–1368.

9 **"One cannot think well":** Virginia Woolf, *A Room of One's Own* (New York: Harcourt, Brace, 1929).

CHAPTER 1: A TASTE OF THINGS TO COME

14 **There were religious sites with temples:** I am indebted to Vanderbilt University archaeologists Steve Werneke and Tiffiny Tung for guiding me through research on the ancient history of agriculture.

15 **"We did not domesticate":** Yuval Noah Harari, *Sapiens: A Brief History of Humankind* (New York: HarperCollins, 2015), 81. Harari's *Sapiens* provides fascinating insights into the impact of farming on early human settlements. Harari adds: "When humans learned to farm in the Agricultural Revolution, their collective power to shape their environment increased, but the lot of many individual humans grew harsher. Peasants had to work harder than foragers to eke out less varied and nutritious food, and they were far more exposed to disease and exploitation." Ibid, 377.

15 **Bioarchaeologists have found lesions:** Lori E. Wright and Francisco Chew, "Porotic Hyperostosis and Paleoepidemiology: A Forensic Perspective on Anemia Among the Ancient Maya," *American Anthropologist* 100 (1998): 924–939.

15 **"Every new agricultural tool":** Personal communication, Ruth DeFries, Sept. 2018. I owe a huge debt to DeFries for the insights and inspiration in her book *The Big Ratchet,* which explores the technologies and innovations that have transformed agriculture since early civilization. Ruth DeFries, *The Big Ratchet: How Humanity Thrives in the Face of Natural Crisis* (New York: Basic Books, 2014).

17 **the Pentagon warned:** Coral Davenport, "Pentagon Signals Security Risks of Climate Change," *New York Times,* Oct. 13, 2014.

17 **helped foment the Arab Spring:** C. E. Werrell and F. Femia, "The Arab Spring and Climate Change," Center for Climate and Security, 2013.

18 **"The power of population":** Thomas Malthus, *An Essay on the Principle of Population* (London: St. Paul's ChurchYard, 1798).

18 **chemists discovered:** DeFries, *The Big Ratchet.*

19 **global food supply jumped 200 percent:** Tim Dyson, "World Food Trends and Prospects to 2025," *Proceedings of the National Academy of Sciences of the United States of America* 96 (1999): 5929–5936.

19 **The world's population more than doubled:** This widely cited fact is well explained with graphics and animation at worldpopulationhistory.org and https://ourworldindata.org/world-population-growth.

19 **less than half its total weight:** Personal communication, Rosemarie Nold, a professor in the Department of Animal Science at South Dakota State University, Dec. 2018.

19 **"producing more food":** Paul Roberts, *The End of Food: The Coming Crisis in the World Food Industry* (London: Bloomsberry, 2008), xi.

20 **Farming has altered the natural systems:** Many texts have explored the way agriculture has transformed natural systems over time, but I particularly like Elizabeth Kolbert's reporting on this topic in the chapter called "Welcome to the Anthropocene" in her book, *The Sixth Extinction* (New York: Henry Holt, 2014).

20 **five billion cattle:** "Live Animals," FAO database, https://tinyurl.com/y9242tar.

20 **"It's tempting to idealize":** Personal communication, Ruth DeFries, Sept. 2018.

20 **More than 800 million:** "The State of Food Security and Nutrition in the World," FAO, 2018, https://tinyurl.com/y8jfvy9o.

21 **a third of the food:** "Food Loss and Waste Facts," FAO, July 22, 2015, https://tinyurl.com/y8twh8nm.

21 **algae blooms that suffocate:** Allen G. Good and Perrin H. Beatty, "Fertilizing Nature: A Tragedy of Excess in the Commons," *PLOS Biology* 9 (2011): 8.

21 **Herbicides and fungicides have squelched:** Johann G. Zaller, Florian Heigl, Liliane Reuss, and Andrea Grabmeir, "Glyphosate Herbicide Affects Belowground Interactions Between Earthworms and Symbiotic Mycorrhizal Fungi in a Model Ecosystem," *Scientific Reports* 4 (2014).

21 **Pesticides have caused mass die-offs:** Chensheng Lu, Kenneth M. Warchol, Richard A. Callahan, et al., "Sub-Lethal Exposure to Neonicotinoids Impaired Honey Bees Winterization Before Proceeding to Colony Collapse Disorder," *Bulletin of Insectology* 67 (2014): 125–130. See also "EPA Actions

to Protect Pollinators," Environmental Protection Agency, https://tinyurl.com/yc6jsceg.

21 **between 1960 and 2000:** Jorge Fernandez-Cornejo et al.,"Pesticide Use in U.S. Agriculture: 21 Selected Crops, 1960–2008," U.S. Department of Agriculture, May 2014, https://tinyurl.com/y7qg42aq.

22 **a decline in nutrition:** D. R. Davis, M. D. Epp, and H. D. Riordan, "Changes in USDA Food Composition Data for 43 Garden Crops, 1950 to 1999," *Journal of the American College of Nutrition* (Dec. 23, 2004): 669–682.

22 **Average sugar consumption:** Michael Moss, *Salt, Sugar, Fat* (New York: Random House, 2013).

22 **diabetes has increased 700 percent:** Gary Taubes, "Why Nutrition Is So Confusing," *New York Times,* Feb. 8, 2014.

22 **serious levels of both undernutrition and obesity:** "Double Burden of Malnutrition," World Health Organization, https://tinyurl.com/y7mwlzy6.

23 **"No one has yet discovered":** Bee Wilson, "The Last Bite," *The New Yorker,* May 19, 2008.

24 **deep distrust of technology as applied to food:** Calestous Juma, a professor at Harvard's Kennedy School of Government, wrote a valuable book on the fear of technology as it relates to many aspects of culture, from agriculture to music: *Innovation and Its Enemies: Why People Resist New Technologies* (New York: Oxford University Press, 2016).

24 **fourfold increase in breast cancer:** Barbara A. Cohn et al., "DDT Exposure in Utero and Breast Cancer," *Journal of Clinical Endocrinology & Metabolism* 100 (2015): 2865–2872.

24 **Roundup was deemed a threat:** Daniel Cressey, "Widely Used Herbicide Linked to Cancer" *Scientific American,* Mar. 25, 2015, https://tinyurl.com/y7ej4c43.

24 **linked to hyperactivity disorder:** D. McCann et al., "Food Additives and Hyperactive Behaviour in 3-Year-Old and 8/9-Year-Old Children in the Community: A Randomised, Double-Blinded, Placebo-Controlled Trial," *Lancet* 370 (2007): 1560–1567. See also Rebecca Harrington, "Does Artificial Food Coloring Contribute to ADHD in Children?" *Scientific American,* Apr. 27, 2015, https://tinyurl.com/h57lv4z.

24 **"has fallen short of the promise":** Danny Hakim, "Doubts About the Promised Bounty of Genetically Modified Crops," *New York Times,* Oct. 29, 2016.

24 **Next Green Revolution:** Tim Folger, "The Next Green Revolution," *National Geographic,* Sept. 2013.

25 **The rift between:** The journalist Charles Mann has written an excellent historical investigation of this rift: *The Wizard and the Prophet* (New York: Alfred A. Knopf, 2018).

26 **a manifesto he's published:** Chris Newman, "Clean Food: If You Want to Save the World, Get Over Yourself," *Medium*, Jan. 28, 2018, https://tinyurl.com/y7nsbyy4.

CHAPTER 2: KILLING FIELDS

30 **"No orchard's the worse":** Robert Frost, "Good-bye, and Keep Cold," in *The Poetry of Robert Frost*, ed. Edward Connery Lathem (New York: Holt Paperbacks, 1979).

34 **1000 BC, the seeds of apple trees:** Naibin Duan, Yang Bai et al., "Genome Re-sequencing Reveals the History of Apple and Supports a Two-Stage Model for Fruit Enlargement," *Nature Communications* 8 (Aug. 15, 2017): 249. Michael Pollan also offers a vivid history of apple tree domestication in *The Botany of Desire: A Plant's Eye View of the World* (New York: Random House, 2002).

34 **"spirited and racy":** H. D. Thoreau, "Wild Apples," *The Atlantic Monthly* 10 (Nov. 1862), https://tinyurl.com/y8oavrpu.

34 **Apples are heterozygous:** Rebecca Rupp, "The History of the 'Forbid den' Fruit," *National Geographic*, July 2014.

34 **grafted on to a scion or rootstock:** Ed Yowell, "Our Disappearing Apples," *The Atlantic*, Nov. 22, 2010.

35 **More pesticides and fungicides:** "2017 Agricultural Chemical Use: Fruit Crops," Agricultural Chemical Use Program of USDA's National Agricultural Statistics Service (July 2018).

35 **six to twelve months:** "Just How Old Are the 'Fresh' Fruit and Vegetables We Eat?" *Guardian*, July 13, 2003.

35 **little to no decline of antioxidant activity:** Addie A. van der Sluis et al., "Polyphenolic Antioxidants in Apples. Effect of Storage Conditions on Four Cultivars," *Acta horticulturae* 600 (Mar. 2003).

35 **40 percent of an apple's antioxidants decline:** Andrea Tarozzi, Alessandra Marchesi, Giorgio Cantelli-Forti, and Patrizia Hrelia, "Cold-Storage Affects Antioxidant Properties of Apples in Caco-2 Cells," *Journal of Nutrition* 134, no. 5 (May 1, 2004): 1105–1109.

35 **$4 billion industry:** "Apple Industry at-a-Glance," U.S. Apple Association, https://tinyurl.com/ya7x5tnh.

35 **comes from China:** "Crops (2016): Apple Quantity Production by Country," FAO database, https://tinyurl.com/l345lur.

37 **6 million in 1910:** E. Dana Durand and William. J. Harris, "Chapter 1: Farms and Farm Property," in *Agriculture 1909 and 1910*, Department of Commerce, Bureau of the Census, https://tinyurl.com/ycrp4q94.

37 **2 million today:** "2012 Census Volume 1, Chapter 1: U.S. National Level Data," USDA, 2012, https://tinyurl.com/jm2u4xe.

37 **"Those who labor in the earth":** Thomas Jefferson, *Notes on the State of Virginia* (London: John Stockdale, 1787), 208.

37 **average age of a farmer:** USDA, 2012.

37 **1 percent of our population:** "Employment by Major Industry Sector," Bureau of Labor Statistics, 2016, https://tinyurl.com/ycecorbf.

37 **40 percent of our land:** "2012 Census of Agriculture: Farms and Farmland," USDA, 2014, https://tinyurl.com/y97c58f3.

42 **800 to 1800 units:** I am grateful to Amaya Atucha, a fruit horticulturist at the University of Wisconsin–Madison, for guiding me through the science behind the life cycle, chilling needs, and blooming trends of fruit trees in the Upper Midwest. The UW Fruit Program website provides an excellent resource for information about critical temperatures for tree fruit bud stages, https://fruit.wisc.edu/. Dr. Greg Lang, a professor of horticulture science at Michigan State University, and Dr. James Luby, who teaches the same at the University of Minnesota, were also instrumental in explaining the impacts of climate change on fruit tree production in the Upper Midwest.

43 **the Valentine's Day Peach Massacre:** Patrick Farrell, "Yes, We Have No Peaches," *New York Times*, Aug. 1, 2016.

44 **rejecting climate science:** Oliver Milman, "Donald Trump Would Be World's Only National Leader to Reject Climate Science," *Guardian*, July 12, 2016.

44 **A map of the states:** "Presidential Election Results: Donald J. Trump Wins," *New York Times*, Aug. 9, 2017, https://tinyurl.com/kvkqlfq.

46 **first released in 2016:** Jeff Andresen, "Climate Change in the Great Lakes Region," Great Lakes Integrated Sciences Assessments, July 2014, http://glisa.umich.edu/climate.

47 **selecting for traits from "low-chill" peach trees:** Details on Dr. José Chaparro's peach-breeding research can be found online at the University of Florida Horticultural Sciences Department, https://tinyurl.com/y94qfl3t.

47 **more than 1.25 inches of rain fell:** Dr. Jerry Hatfield guided me through the details of his research, an overview of which can be found at https://tinyurl.com/y74qnqd8. Hatfield further explains his findings in a University of Minnesota presentation, "Climate Change Affecting Agriculture," https://tinyurl.com/ya4x8urc.

48 **plants have to cool themselves:** Anthony W. King, Carla A. Gunderson, Wilfred M. Post, David J. Weston, and Stan D. Wullschleger, "Plant Respiration in a Warmer World," *Science* 312 (Apr. 28, 2006): 536–537.

48 **the "New Dust Bowl":** Laura Parker, "Parched: A New Dust Bowl Forms in the Heartland," *National Geographic*, May 2014.

48 **Crop insurance payouts totaled about $30 billion:** "EWG's Farm Subsidy Database: Crop Insurance," *Environmental Working Group,* https://tinyurl.com/y776kyao.

48 **rise by 3 to 4 degrees:** "Interior Department Releases Report Underscoring Impacts of Climate Change on Western Water Resources," U.S. Department of the Interior, Mar. 22, 2016.

48 **sharp decline in the snowpack:** Ibid.

48 **"We're really just beginning to grasp":** Margaret Walsh led one of the first comprehensive studies produced on the impacts of warming trends on food production; see "The Effects of Climate Change on Agriculture, Land Resources, Water Resources, and Biodiversity in the United States," *U.S. Climate Change Science Program Assessment Product* 4.3 (May 2008).

49 **the "fruit basket, the salad bowl":** Matt Black, "The Dry Land," *The New Yorker,* Sept. 22, 2014.

49 **More than half a million acres of crops:** Phillip Reese, "Study: California Farmers to Fallow 560,000 Acres of Crops This Year," *Sacramento Bee,* July 2, 2015. See also this Pulitzer-winning article on the impacts of these fallowed fields on California farmers: Diana Marcum, "Scenes from California's Dust Bowl," *Los Angeles Times,* Dec. 10, 2014.

49 **feeding their bees sugar syrup:** Ezra David Romero, "Drought Is Driving Beekeepers and Their Hives from California," *The Salt,* NPR, Sept. 29, 2015.

50 **thermal shock:** Dr. Jason Londo, a scientist in the Grape Genetics Research Unit at the USDA, explained the impacts of climate change on orchards and viticulture. See also Jason Londo, "Characterization of Wild North American Grapevine Cold Hardiness Using Differential Thermal Analysis," *American Journal of Enology and Viticulture* 68, no. 2 (2017): 203–212.

50 **avocado farms had begun to suffer:** Ruth Tam, "Guacamole at Chipotle Could Be Climate Change's Next Casualty," *PBS News Hour,* Mar. 4, 2014.

50 **hurting the olive fields of Trevi:** Somini Sengupta, "How Climate Change Is Playing Havoc with Olive Oil (and Farmers)," *New York Times,* Oct. 24, 2017.

50 **Heat was taking a toll:** Eric Randolph,"Iran's Pistachio Farms Are Dying of Thirst," Phys.org, Sept. 4, 2016, https://tinyurl.com/y8rzp9pb.

50 **chocolate was coming under threat:** Michon Scott, "Climate and Chocolate," National Oceanic and Atmospheric Administration, Feb. 10, 2016.

50 **bearing down on the coffee farms:** Corey Watts, "A Brewing Storm: The Climate Change Risks to Coffee," Climate Institute, 2016.

51 **cut in half:** Ibid., 6.

51 **developing a range of responses:** "Annual Report 2017: Creating the Future of Coffee," World Coffee Research, 2017.

53 **frost fans on orchards in Michigan:** Ross Courtney, "Arctic Armor: Methods of Combating Frost," *Good Fruit Grower,* Sept. 6, 2017.

54 **development of "cryoprotectants":** Alabama Extension System, "Methods of Freeze Protection for Fruit Crops," Alabama A&M and Auburn Universities, 2000.

54 **formulating "growth regulators":** Amy Irish-Brown and Phil Schwallier, "Setting Apples with Plant Growth Regulators," Michigan State University Extension, May 9, 2017.

CHAPTER 3: SEEDS OF DROUGHT

58 **half of the calories consumed nationwide:** Haradhan Kumar Mohajan, "Food and Nutrition Scenario of Kenya," *American Journal of Food and Nutrition* 2 (2014): 28–38.

64 **worst droughts on record:** "Ethiopia Crisis," U.S. Agency for International Development, 2017, https://tinyurl.com/yavkfo7w.

64 **about twelve million people:** "U.N. Calls for More Funds to Save Lives Across Horn of Africa," UN News, July 29, 2011.

64 **Ten million were receiving emergency food aid:** Michael Klaus, "UNICEF Responds to Horn of Africa Food Crisis That Has Left 2 Million Children Malnourished," UNICEF, July 11, 2011.

65 **2.4 million Kenyans:** "Country Brief: Kenya," FAO, May 8, 2018, https://tinyurl.com/y7qq72j8.

65 **The glaciers on Mount Kenya:** "Kenya: Atlas of Our Changing Environment," United Nations Environment Programme, 2009. See also Rupi Managat, "The Vanishing Glaciers of Mt. Kenya," *The East African,* Jan. 14, 2017.

65 **three million camels in Kenya:** Andrea Dijkstra, "Kenyans Turn to Camels to Cope with Climate Change," Deutsche Welle, Apr. 24, 2017.

66 **"Most farmers in Kenya":** Personal communication, Robb Fraley, Jan. 2016.

67 **DroughtTego seeds are gene-altered:** I'm indebted to Pamela Ronald, plant scientist at UC Davis, for her detailed explanations of the distinction between genetically *engineered* and genetically *modified* plants. While Tego seeds are developed using genome analysis and marker-assisted breeding techniques, they are not modified with a gene from a different plant or animal species. Ronald describes the virtues of both genetic engineering and modification in the book she wrote with her husband, Raoul Adamchak, an organic farmer: *Tomorrow's Table: Organic Farming, Genetics, and the Future of Food* (New York: Oxford University Press, 2008).

67 **The government here banned:** K. Snipes and C. Kamau, "Kenya Bans Genetically Modified Food Imports," USDA Foreign Agricultural Service, Nov. 2012.

67 **Of the fifty-four countries:** Steven Cerier, "Led by Nigeria, Africa Opening Door to Genetically Modified Crop Cultivation," Genetic Literacy Project, Mar. 6, 2017, https://tinyurl.com/ycl5cx63.

67 **a growing number of Western:** Tamar Haspel, "The Last Thing Africa Needs to Be Debating Is GMOs," *Washington Post,* May 22, 2015, https://tinyurl.com/yaw9sx04. See also MacDonald Dzirutwe, "Africa Takes Fresh Look at GMO Crops as Drought Blights Continent," Reuters, Jan. 7, 2016, https://tinyurl.com/y7sqsswm.

69 **early evolution of wheat and barley:** Jared Diamond, "Evolution, Consequences and Future of Plant and Animal Domestication," *Nature* 418 (2002): 700–707.

69 **George Beadle discovered:** Sean B. Carroll, "Tracking the Ancestry of Corn Back 9,000 Years," *New York Times,* May 24, 2010.

70 **fifteen million calories per acre:** Tamar Haspel, "In Defense of Corn, the World's Most Important Food Crop," *Washington Post,* July 12, 2015.

71 **genes from the cecropia moth:** E. E. Borejsza-Wysocka, M. Malnoy, H. S. Aldwinckle, S. V. Beer, J. L. Norelli, and S. H. He, "Strategies for Obtaining Fire Blight Resistance in Apple by rDNA Technology," *Acta horticulturae* 738 (2007): 283–285.

71 **DroughtTela, the GMO version:** Personal communication, Fraley.

71 **90 percent of the corn:** "Recent Trends in GE Adoption," U.S. Department of Agriculture, 2018.

71 **"The company has donated":** Personal communication, Fraley.

73 **Seventy percent of processed American foods:** Alison Moodie, "GMO Food Labels Are Coming to More US Grocery Shelves—Are Consumers Ready?" *Guardian,* Mar. 24, 2016.

73 **400 million acres of GMO crops:** "Global Status of Commercialized Biotech/GM Crops: 2017," International Service for the Acquisition of Agri-biotech Applications (ISAAA), 2018.

73 **about 180 million:** Ibid., 6.

73 **"Manufacturers of genetically altered foods":** Martha R. Herbert, "Feasting on the Unknown," *Chicago Tribune,* Sept. 3, 2000.

74 **only major scientific study to connect GMOs to cancer:** Gilles-Eric Séralini, Emilie Clair, Robin Mesnage, Steeve Gress, Nicolas Defarge, Manuela Malatesta, Didier Hennequin, and Joël Spirouxde Vendômois, "Long Term Toxicity of a Roundup Herbicide and a Roundup-Tolerant Genetically Modified Maize," *Food and Chemical Toxicology* 50 (2012): 4221–4231.

75 **he published a mea-culpa book:** Mark Lynas, *Seeds of Science: Why We Got It So Wrong on GMOs* (New York: Bloomsbury Sigma, 2018).

75 **"The argument against GMOs":** Tamar Haspel, "The Public Doesn't Trust GMOs: Will It Trust CRISPR?" *Vox,* July 26, 2018.

75 **scientists began engineering enzymes:** Siddhartha Mukherjee offers an interesting, accessible history of gene editing in *The Gene: An Intimate History* (New York: Scribner, 2016). For a deeper dive into the science, see also Susan Aldridge, *The Thread of Life: The Story of Genes and Genetic Engineering* (Cambridge: Cambridge University Press, 1996).

75 **scientists used CRISPR to erase HIV:** Alice Park, "HIV Genes Have Been Cut Out of Live Animals Using CRISPR," *Time,* May 19, 2016.

75 **piglets with DNA scrubbed of a lethal virus:** Kelly Servick, "Gene-Editing Method Revives Hopes for Transplanting Pig Organs into People," *Science,* Oct. 11, 2015.

77 **40 percent of Americans:** Carrie Funk and Brian Kennedy, "The New Food Fights: U.S. Public Divides over Food Science," Pew Research Center, Dec. 1. 2016.

77 **"The companies that produce [non-GMO] brands":** Michael Gerson, "Are You Anti-GMO? Then You're Anti-science, Too," *Washington Post,* May 3, 2018.

77 **a tenfold reduction:** Jorge Fernandez-Cornejo, Seth Wechsler, Mike Livingston, and Lorraine Mitchell, "Genetically Engineered Crops in the United States," U.S. Department of Agriculture, Feb. 2014.

77 **papayas genetically engineered to resist ringspot virus:** Carol Kaesuk Yoon, "Stalked by Deadly Virus, Papaya Lives to Breed Again," *New York Times,* July 20, 1999.

77 **"scuba rice":** Drake Baer, "Bill Gates Is Betting on a Strain of Rice That Can Survive Floods," *Business Insider,* Sept. 2, 2015.

78 **An op-ed by a Zimbabwean scientist:** Nyasha Mudukuti, "We May Starve, but at Least We'll Be GMO-Free," *Wall Street Journal,* Mar. 10, 2016.

82 **"yields about 40 percent higher":** Personal communication, Dr. Dickson Liyago, July 2016.

82 **"transgenic corn pollen is harmless":** Tom Clarke, "Monarchs Safe from Bt," *Nature,* Sept. 12, 2001.

84 **95 percent of its water:** J. M. Farrant, K. Cooper, A. Hilgart, K .O. Abdalla, J. Bentley, J. A. Thomson, H .J. Dace, N. Peton, S. G. Mundree, and M. S. Rafudeen, "A Molecular Physiological Review of Vegetative Desiccation Tolerance in the Resurrection Plant *Xerophyta viscosa,*" *Planta* 242 (2015): 407–426.

84 **"resurrection" genes into tobacco:** David Shamah, "'Hibernating' Crops May Be Science's Cure for Drought," *Times of Israel,* Aug. 29, 2013.

84 **a soy plant spliced with genes:** Liliana Samuel, "Drought-Resistant Argentine Soy Raises Hopes, Concerns," Phys.org, Apr. 27, 2012. See also Lizzie Wade, "Argentina May Have Figured Out How to Get GMOs Right," *Wired*, Oct. 28, 2015.

CHAPTER 4: ROBOCROP

89 **killing an estimated 15,000 people:** Alan Taylor, "Bhopal: The World's Worst Industrial Disaster, 30 Years Later," *The Atlantic*, Dec. 2, 2014, https://tinyurl.com/y967zce9.

89 **castrating male frogs:** Tyrone B. Hayes et al., "Atrazine Induces Complete Feminization and Chemical Castration in Male African Clawed Frogs *(Xenopus laevis)*," *Proceedings of the National Academy of Sciences of the United States of America* 107, no. 10 (Mar. 9, 2010): 4612–4617.

89 **more than 8000 square miles:** "Gulf of Mexico 'Dead Zone' Is the Largest Ever Measured" (media release), National Oceanic and Atmospheric Administration, Aug. 2, 2017, https://bit.ly/2vtOOnF.

89 **Those chemical fertilizers:** Ly Truong and Claire Press, "Making Food Crops That Feed Themselves," BBC, June 8, 2018, https://tinyurl.com/yafaelba.

90 **"blue baby syndrome":** Clay Masters, "Iowa's Nasty Water War," *Politico*, Jan. 21, 2016, https://tinyurl.com/y7vcq6mj.

90 **A number of recent academic studies:** Ying Zhang et al., "Prenatal Exposure to Organophosphate Pesticides and Neurobehavioral Development of Neonates: A Birth Cohort Study in Shenyang, China," *PLOS ONE* (2014). See also Stephen A. Rauch et al., "Associations of Prenatal Exposure to Organophosphate Pesticide Metabolites with Gestational Age and Birth Weight," *Environmental Health Perspectives* 120 (2012): 1055–1060.

90 **according to most toxicology research:** "Pesticide Residues in Food," World Health Organization, Feb. 19, 2018. See also Lois Swirsky Gold et al., "Pesticide Residues in Food and Cancer Risk: A Critical Analysis," in *Handbook of Pesticide Toxicology*, ed. R. Krieger, 2d ed. (San Diego: Academic Press, 2001), 799–843.

93 **By early 2017, about a fifth of all the lettuce:** Personal communication, Jorge Heraud, Jan. 2018.

96 **elegant masters of adaptation:** Richard Mabey, *Weeds: In Defense of Nature's Most Unloved Plants* (New York: Ecco/HarperCollins, 2011).

97 **the Genghis Khan of weeds:** Travis Legleiter and Bill Johnson, "Palmer Amaranth Biology, Identification, and Management," Purdue Extension, Nov. 2013, https://tinyurl.com/ycqqlncj. See also Brooke Borel, "Weeds Are Winning the War Against Herbicide Resistance," *Scientific American*, June 18, 2018, https://tinyurl.com/yczrj2t8.

97 **an Arkansas cotton farmer:** Tom Barber, professor of weed science for the Crop, Soil and Environmental Science Department at the University of Arkansas System Division of Agriculture, generously explained the history and science of pigweed and Roundup resistance. See also Ryan McGeeney, "As Arkansas Growers Struggle with Increasingly Resistant Weeds, State Weighs Labeling," University of Alabama Division of Agriculture Research and Extension, Nov. 21, 2016.

97 **herbicide-resistant weeds infest 70 million acres of crops:** Carey Gillam, "EPA Approves Dow's Enlist Herbicide for GMOs," *Scientific American*, 2014, https://tinyurl.com/ycrckvvq.

98 **millions of acres of neighboring crops:** Eric Lipton, "Crops in 25 States Damaged by Unintended Drift of Weed Killer," *New York Times*, Nov. 1, 2017.

98 **one disagreement ended in murder:** Marianne McCune, "A Pesticide, a Pigweed and a Farmer's Murder," NPR, June 14, 2017, https://tinyurl.com/y8yvhy7a.

102 **lost a third of its arable soil:** "Status of the World's Soil Resources," FAO 2015, https://tinyurl.com/ya07z4ue. See also F. Nachtergaele, R. Biancalani, and M. Petri, "Land Degradation: SOLAW Background Thematic Report 3," FAO, 2011, https://tinyurl.com/y8ysv43x.

102 **5.6 billion pounds:** Michael C. R. Alvanja, "Pesticides Use and Exposure Extensive Worldwide," *Reviews on Environmental Health* 24, no. 4 (2009): 303–309.

102 **the product of biological weapons research:** Daniel Charles, *Master Mind: The Rise and Fall of Fritz Haber, the Nobel Laureate Who Launched the Age of Chemical Warfare* (New York: Ecco/HarperCollins, 2005). See also Jad Abumrad and Robert Krulwich's episode of *Radiolab* titled "How Do You Solve a Problem Like Fritz Haber?" in which they grapple with the good and evil that has come from Haber's work, WNYC, 2012, https://tinyurl.com/y7hq7hyg.

102 **shot up more than fifteenfold:** Brett Cherry, "GM Crops Increase Herbicide Use in the United States," *Science in Society* 45 (2010): 44–46.

103 **increased by 500 percent:** Paul J. Mills, Izabela Kania-Korwel, John Fagan, Linda K. McEvoy, Gail A. Laughlin, and Elizabeth Barrett-Connor, "Excretion of the Herbicide Glyphosate in Older Adults Between 1993 and 2016," *Journal of the American Medical Association* 318, no. 16 (2017): 1610–1611.

103 **glyphosate in high levels is a "probable carcinogen":** "Glyphosate Issue Paper: Evaluation of Carcinogenic Potential," Environmental Protection Agency Office of Pesticide Programs, Sept. 12, 2016. See also L. N. Vandenberg, B. Blumberg, M. N. Antoniou et al., "Is It Time to Reassess Current Safety Standards for Glyphosate-Based Herbicides?" *Journal of Epidemiology and Community Health* 71, no. 6 (2017): 613–618.

103 **Recent studies have also tied high levels:** Dandan Yan, Yunjian Zhang, Liegang Liu, and Hong Yan, "Pesticide Exposure and Risk of Alzheimer's Disease: A Systematic Review and Meta-analysis," *Scientific Reports* 6 (2016). See also Jane A. Hoppin et al., "Pesticides Are Associated with Allergic and Non-Allergic Wheeze Among Male Farmers," *Environmental Health Perspectives* 125, no. 4 (Apr. 2017): 535–543; and Marys F. Bouchard et al., "Attention-Deficit/Hyperactivity Disorder and Urinary Metabolites of Organophosphate Pesticides," *Pediatrics* 125, no. 6 (June 2010): 1270–1277.

103 **harm soil microbiology:** Johann G. Zaller, Florian Heigl, Liliane Ruess, and Andrea Grabmaier, "Glyphosate Herbicide Affects Belowground Interactions Between Earthworms and Symbiotic Mycorrhizal Fungi in a Model Ecosystem," *Scientific Reports* 4 (2014).

103 **overstimulate soil microbes:** Christopher Ratzke, Jonas Sebastian Denk, and Jeff Gore, "Ecological Suicide in Microbes," *Nature Ecology and Evolution* 2 (2018): 867–872.

103 **degrading ten times faster than they can be replenished:** S. Lang, "'Slow, Insidious' Soil Erosion Threatens Human Health and Welfare as Well as the Environment, Cornell Study Asserts," *Cornell Chronicle*, Mar. 20, 2006, https://tinyurl.com/y7j928jr.

104 **herbicides that first enabled the no-till practice:** David R. Huggins and John P. Reganold, "No-Till: The Quiet Revolution," *Scientific American*, July 2008.

104 **a fifth of U.S. cropland:** Elizabeth Creech, "Saving Money, Time and Soil: The Economics of No-Till Agriculture Farming," U.S. Department of Agriculture, Nov. 30, 2017, https://tinyurl.com/yadsfvjn. See also A. Kassam, T. Friedrich, R. Derpsch, and J. Kienzle, "Overview of the Worldwide Spread of Conservation Agriculture," *Field Actions Science Reports* 8 (2015), https://tinyurl.com/y8q70xmk.

105 **billions of microbes:** Elaine R. Ingham, "Soil Bacteria," USDA Natural Resource Conservation Service, https://tinyurl.com/y8u6sk9u. See also Mike Amaranthus and Bruce Allyn, "Healthy Soil Microbes, Healthy People," *The Atlantic*, June 11, 2013, https://tinyurl.com/ydgf300x.

105 **"protect the health of the soil":** See also David R. Montgomery, *Dirt: The Erosion of Civilizations* (Berkeley: University of California Press, 2007). Montgomery demonstrates the way poor soil stewardship has time and time again undermined the success of civilizations since the dawn of humankind.

105 **hundreds of millions of dollars:** Tom Simonite, "Why John Deere Just Spent $305 Million on a Lettuce Farming Robot," *Wired*, Sept. 6, 2017, https://tinyurl.com/ycnk03r2.

107 **right-to-repair movement:** "A 'Right to Repair' Movement Tools Up," *The Economist*, Sept. 30, 2017.

CHAPTER 5: SENSOR SENSIBILITY

111 **"Tech should serve humanity":** Quoted in Nanette Byers, "Tim Cook: Technology Should Serve Humanity, Not the Other Way Around," *MIT Technology Review,* June 9, 2017.

111 **world's biggest food producer:** "Production Indices: Visualize Data," FAO database, https://tinyurl.com/y725a4vg.

111 **an article in *Caixin*:** Ma Yuan, Zheng Fei, Liu Ran, and Rong Tiankun, "Hedge Funds Bet on Organic Farming in China," *Caixin* online, Mar. 14, 2013, https://tinyurl.com/y8uc5xaq.

112 **lost crucial farmland:** Kaifang Shi et al., "Urban Expansion and Agricultural Land Loss in China: A Multiscale Perspective," *Sustainability* 8 (2016), https://tinyurl.com/ydytatub.

112 **half of their food from local farms:** Personal communication, Junshi Chen, May 2014.

112 **deemed his city of 24 million people "unlivable":** Brook Larmer, "How Do You Keep Your Kids Healthy in Smog-Choked China?" *New York Times,* Apr. 16, 2015.

112 **WHO's maximum safe level:** Didi Kirsten Tatlow, "Don't Call It 'Smog' in Beijing, Call It a 'Meteorological Disaster,'" *New York Times,* Dec. 15, 2016.

112 **a quarter of the country's lakes and rivers:** "2007 Water Resources of the Yellow River Bulletin," Ministry of Water Resources, Yellow River Conservancy Commission, https://tinyurl.com/ya6hdj28. See also David Stanway, "Pollution Makes Quarter of China Water Unusable: Ministry," Reuters, July 26, 2010, https://tinyurl.com/y8ohrfgw.

112 **thirty fastest-growing cities have "medium to high":** Daniel Shemie, Kari Vigerstol, Mu Quan, and Wang Longzhu, "China's New Opportunity: Water Funds," The Nature Conservancy, 2016, https://tinyurl.com/ya9bbuqa.

112 **20 percent of the country's farmland:** "The Bad Earth: The Most Neglected Threat to Public Health in China Is Toxic Soil," *The Economist,* June 8, 2017, https://tinyurl.com/ybt976kp.

113 **the government created a special police unit:** Personal communication, Chen. See also John Balzano, "The Food Police: China Proposes a Plan for a Special Unit for Food and Drug Safety Violations," *Forbes,* Apr. 20, 2014.

113 **arsenic in apple juice:** Karlynn Fronek, "Concerns Surrounding Imported Fruit Juice from China," *Food Quality and Safety,* June 12, 2014, https://tinyurl.com/y77e3fwt.

113 **authorities have executed two workers:** Tania Branigan, "China Executes Two for Tainted Milk Scandal," *Guardian,* Nov. 24, 2009.

113 **"Two hundred million Chinese every year"**: Personal communication, Chen, May 2014. See also Yanzhong Huang, "China's Worsening Food Safety Crisis," *The Atlantic*, Aug. 28, 2012.

113 **200 million farms in China**: Tracy McMillan, "How China Plans to Feed 1.4 Billion Growing Appetites," *National Geographic*, Feb. 2018. See also "Employment in Agriculture (% of Total Employment)," World Bank Database, retrieved Sept. 2018, https://tinyurl.com/y7v6y2by.

118 **more than a quarter of the rice**: "The Bad Earth: The Most Neglected Threat to Public Health," 2017.

119 **100 cubic meters of water per year**: "Thirsty Beijing to Raise Water Prices in Conservation Push," Reuters, Apr. 29, 2014.

119 **2840 cubic meters U.S. citizens consume**: Mark Fischetti, "How Much Water Do Nations Consume?" *Scientific American*, May 21, 2012.

121 **a fifth of the country's farmland was contaminated**: Scott Neuman, "China Admits That One-Fifth of Its Farmland Is Contaminated," NPR, Apr. 18, 2014.

121 **In 2016, chemical residues**: Javier C. Hernandez, "Chinese Parents Outraged After Illnesses at School Are Tied to Pollution," *New York Times*, Apr. 18, 2016.

121 **hepatitis A, typhoid, and certain cancers**: Yonglong Lu, Shuai Song, Ruoshi Wang, Zhaoyang Liu, Jing Meng, Andrew J. Sweetman, Alan Jenkins, Robert C. Ferrier, Hong Li, Wei Luo, and Tieyu Wang, "Impacts of Soil and Water Pollution on Food Safety and Health Risks in China," *Environment International* 77 (April 2015): 5–15.

121 **issued a plan**: David Stanway, "China to Make More Polluted Land Safer for Agriculture by 2020," Reuters, Feb. 4, 2018, https://tinyurl.com/ycx2g544.

121 **it deployed a regiment**: Samuel Osborne, "China Reassigns 60,000 Soldiers to Plant Trees in Bid to Fight Pollution," *Independent*, Feb. 13, 2018, https://tinyurl.com/yccddz6d.

122 **"phytoremediation is very hard to do effectively"**: Personal communication, Yonglong Lu, Oct. 2018.

122 **crops that won't absorb toxins**: Haiyang Liu, Miao Ren, Jiao Qu, Yue Feng, Xiangmeng Song, Qian Zhang, Qiao Cong, and Xing Yuan, "A Cost-Effective Method for Recycling Carbon and Metals in Plants: Synthesizing Nanomaterials," *Environmental Science: Nano* 2 (2017), https://tinyurl.com/yab4pqal. See also Hillary Rosner, "Turning Genetically Engineered Trees into Toxic Avengers," *New York Times*, Aug. 3, 2004.

122 **$18,000 per acre of soil**: "China Needs Patience to Fight Costly War Against Soil Pollution: Government," Reuters, June 22, 2017, https://tinyurl.com/yc94j9v7.

122 **200,000 acres of heavily contaminated farmland:** "Report Sounds Alarm on Soil Pollution," FAO, May 2, 2018, https://tinyurl.com/ybg9bwxk.

122 **pesticide use has more than doubled:** FAO data on mainland Chinese pesticide use 1991–2016 can be found at https://tinyurl.com/ycstwdre.

124 **with products three times the price:** "China, Peoples Republic of: Organics Report," USDA Foreign Agricultural Service, GAIN Report Number 10046, Oct. 26, 2010, https://tinyurl.com/y9uxxx5v. See also Michelle Winglee, "China's Organics Market, Beset by Obstacles, Is Still Taking Off," *Foreign Policy*, Sept. 21, 2016, https://tinyurl.com/y8ztalne.

126 **surpassed $7 billion:** Personal communication, Jay Wang, Oct. 2018.

126 **Organic certifications more than doubled:** Ibid.

126 **tripled its spending on food safety:** Gail Sullivan, "Wal-Mart to Triple Food Safety Spending in China After Donkey Meat Disaster," *Washington Post*, July 17, 2014.

126 **Kroger recently announced:** Liza Lin and Heather Haddon, "Kroger to Sell Groceries on Alibaba Site in China," *Wall Street Journal*, Aug. 14, 2018.

127 **$200 million into Plenty:** Leanna Garfield, "A Jeff Bezos–Backed Warehouse Farm Startup Is Building 300 Indoor Farms Across China," *Business Insider*, Jan. 23, 2018, https://tinyurl.com/y72nk9zs.

127 **by 2020 he'll have built:** Ibid.

127 **900 million acres of farmland:** "Farms and Land in Farms 2017 Summary," U.S. Department of Agriculture, Feb. 2018, https://tinyurl.com/y7vdhu45.

127 **One of the largest vertical-farming companies:** Andrew Buncombe, "AeroFarms: Work Starts to Build World's Largest Vertical Urban Farm in Newark," *Independent*, Apr. 28, 2015, https://tinyurl.com/y7fp3dko.

CHAPTER 6: ALTITUDE ADJUSTMENT

130 **experimental concept of indoor farming called aeroponics:** Scientists first began growing plants in misted air in the 1920s to study root structure. For a history and overview of the method, see P. Gopinath et al., "Aeroponics Soilless Cultivation System for Vegetable Crops," *Chemical Science Review and Letters* 6, no. 22 (2017): 838–849.

131 **Roman emperor Tiberius:** H. S. Paris and J. Janick, "What the Roman Emperor Tiberius Grew in His Greenhouses," Proceedings of the Ninth EUCARPIA Meeting on Genetics and Breeding of Cucurbitaceae, May 21–24, 2008.

132 **first glassed-in greenhouse structures:** Pamela D. Toler, *Inventions in Architecture: From Stone Walls to Solar Panels* (New York: Cavendish Square Publishing, 2017), ch. 5.

132 **roofs can be seen from space:** Tom Hall, "The Plastic Mosaic You Can See from Space: Spain's Greenhouse Complex," *Bloomberg*, Feb. 20, 2015, https://tinyurl.com/ydhr3852.

132 **criticized for producing thousands of tons:** "A Study Links Temperature and Greenhouses in Almería," Reuters, Oct. 8, 2008, https://tinyurl.com/y7qgyr7n. See also Pablo Campra, Monica Garcia, Yolanda Canton, and Alicia Palacios-Orueta, "Surface Temperature Cooling Trends and Negative Radiative Forcing Due to Landuse Change Toward Greenhouse Farming in Southeastern Spain," *Journal of Geophysical Research Atmospheres* 133, no. D18 (2008), https://tinyurl.com/yakszdc6.

132 **70 percent of the produce:** D. L. V. Martínez, L. J. B. Ureña, F. D. M. Aiz, and A. L. Martínez, "Greenhouse Agriculture in Almería," *Serie Economíca* 27 (2016).

132 **its own greenhouses designed:** Frank Viviano, "This Tiny Country Feeds the World," *National Geographic*, Sept. 2017.

133 **wiped out "fully a third":** Personal communication, David Lobell, June 2016.

133 **China has purchased land in thirty-three countries:** Ana Swanson, "An Incredible Image Shows How Powerful Countries Are Buying Up Much of the World's Land," *Washington Post*, May 21, 2015.

133 **"global land grab":** Leon Kaye, "The Global Land Grab Is the Next Human Rights Challenge for Business," *Guardian*, Sept. 11, 2012.

135 **$130 million in funding:** Crunchbase, Aug. 24, 2018, https://tinyurl.com/y9r83tey.

141 **"eleven times higher yields":** "Comparison of Land, Water, and Energy Requirements of Lettuce Grown Using Hydroponic vs. Conventional Agricultural Methods," *International Journal of Environmental Research and Public Health* 12, no. 6 (June 2015): 6879–6891.

141 **LED lighting efficiency jumped about 50 percent:** "LED Light Bulbs Keep Improving in Efficiency and Quality," U.S. Energy Information Administration, Nov. 4, 2014.

142 **"bottled water of produce":** Personal communication, Tamar Haspel, July 2018. See also Tamar Haspel, "Why Salad Is So Overrated," *Washington Post*, Aug. 23, 2015.

143 **most of them thrown away from rot:** Millicent G. Managa et al., "Impact of Transportation, Storage, and Retail Shelf Conditions on Lettuce Quality and Phytonutrients Losses in the Supply Chain," *Food Science and Nutrition*, July 4, 2018, https://tinyurl.com/y9rfmdmv. See also Jean C. Buzby et al., "Estimated Fresh Produce Shrink and Food Loss in U.S. Supermarkets," *Agriculture* 5 (2015): 626–648.

143 **3.5 gallons of water to cultivate:** M. M. Mekonnen and A. Y. Hoekstra, "The Green, Blue and Grey Water Footprint of Crops and Derived Crop Products," *Hydrology and Earth System Sciences* 15 (2011): 1577–1600.

143 **killed five Americans:** Julia Belluz, "No One Should Die from Eating Salad," *Vox*, June 4, 2018, https://tinyurl.com/ycammwre.

143 ***"Terroir"*:** Amy B. Trubek, *The Taste of Place: A Cultural Journey into Terroir* (Berkeley: University of California Press, 2008).

147 **Saudi Arabia, for example, imports 75 percent:** Hussein Mousa, "Saudi Arabia: Exporter Guide 2017," USDA Foreign Agricultural Service, GAIN Report Number SA1710, Dec. 12, 2017, https://tinyurl.com/y9aq7sgr.

147 **Iraq, Qatar, and the United Arab Emirates:** Zahra Babar and Suzi Mirgani, eds., *Food Security in the Middle East* (Oxford: Oxford University Press, 2014).

147 **9.8 billion by 2050:** "The World Population Prospects: The 2017 Revision," UN Department of Economic and Social Affairs, June 21, 2017.

147 **fourteen of the world's twenty biggest metropolises:** Daniel Hoornweg and Kevin Pope, "Population Predictions for the World's Largest Cities in the 21st Century," *Environment and Urbanization* 29, no. 1 (Apr. 2017): 195–216.

147 **raise global food output by 70 percent:** "How to Feed the World in 2050," FAO, 2009, https://tinyurl.com/5ranufw.

CHAPTER 7: TIPPING THE SCALES

151 **About a dozen sea lice:** Gwynn Guilford, "The Gross Reason You'll Be Paying a Lot More for Salmon This Year," *Quartz*, Jan. 22, 2017, https://tinyurl.com/jlto3k3.

152 **declined by 12 percent:** Personal communication, Alf-Helge Aarskog, Sept. 2017.

152 **1.5 billion Atlantic salmon:** Ibid.

152 **by far the biggest producer:** "The Norwegian Aquaculture Analysis 2017," Ernst and Young, 2018, https://tinyurl.com/ycgx2wgd.

152 **doubled in the last decade:** "The State of World Fisheries and Aquaculture," FAO, 2016, https://tinyurl.com/h607rga.

152 **begun to harm other salmon habitats:** Lisa Crozier, "Impacts of Climate Change on Salmon of the Pacific Northwest," National Oceanic and Atmospheric Administration, Oct. 2016. See also Brian Hines, "California's Fishing Industry Is Drying Up: We Need to Think Big on Climate Change," *Sacramento Bee*, Mar. 27, 2018.

154 **come down 20 percent since 2005:** Personal communication, Aarskog.

154 **"only 2 percent of our food"**: Personal communication, Aarskog. See also "The State of World Fisheries and Aquaculture 2006," FAO, 2007, https://tinyurl.com/y8ndv9ad.

154 **"Industrial aquaculture, especially for salmon"**: Personal communication, Don Staniford, Apr. 2018. See also Daniel Pauly, "Aqualypse Now," *The New Republic,* Sept. 27, 2009, https://tinyurl.com/yczluvsv.

155 **5 percent of the global aquaculture industry**: Personal communication, Aarskog.

155 **will grow at least 35 percent**: Joel K. Bourne Jr., "How to Farm a Better Fish," *National Geographic,* June 2014, https://tinyurl.com/yap558yl.

156 **number of young Chinook salmon**: Ryan Sabalow, "Devastated Salmon Population Likely to Result in Fishing Restrictions," *Sacramento Bee,* Feb. 29, 2016, https://tinyurl.com/yc3xlld5. See also Ryan Sabalow, Dale Kasler, and Philip Reese, "Feds: Winter Salmon Run Nearly Extinguished in California Drought," *Sacramento Bee,* Oct. 28, 2015; and Megan Nguyen, "State of the Salmonids II: Fish in Hot Water," UC Davis Center for Watershed Science, May 16, 2017, https://tinyurl.com/ycd9vpgy.

156 **"two-thirds of marine species"**: Erica Goode, "Fish Seek Cooler Waters, Leaving Some Fishermen's Nets Empty," *New York Times,* Dec. 30, 2016.

156 **Lobster populations**: Emily Greenhalgh, "Climate and Lobsters," National Oceanic and Atmospheric Administration, Oct. 6, 2016, https://tinyurl.com/y73m9yxy.

156 **cod fishery in New England**: "New Study: Warming Waters a Major Factor in the Collapse of New England Cod," Gulf of Maine Research Institute, Oct. 29, 2015, https://tinyurl.com/ybwpgkdm.

157 **Black sea bass, scupp, yellowtail flounder**: James W. Morley et al., "Projecting Shifts in Thermal Habitat for 686 Species on the North American Continental Shelf," *PLOS ONE* 13, no. 5 (2018), https://tinyurl.com/y9t9mvw7. See also Shelley Dawicki, "Many Young Fish Moving North with Adults as Climate Changes," National Oceanic and Atmospheric Administration—Northeast Fisheries Science Center, Oct. 1, 2015, https://tinyurl.com/y9gf5kjn.

157 **"The center of the black sea bass population"**: Goode, "Fish Seek Cooler Waters, Leaving Some Fishermen's Nets Empty."

157 **"evil twin of global warming"**: Alex Rogers, "Global Warming's Evil Twin: Ocean Acidification," *The Conversation,* Oct. 9, 2013, https://tinyurl.com/ycusw6la.

157 **Dungeness crab population**: Hal Bernton, "Study Predicts Decline in Dungeness Crab from Ocean Acidification," *Seattle Times,* Jan. 15, 2017, https://tinyurl.com/ydamq9k7. See also "Ocean Acidification Puts NW

Dungeness Crab at Risk: Study Finds Lower pH Reduces Survival of Crab Larvae," *ScienceDaily*, May 18, 2016, https://tinyurl.com/jekd7k6.

157 **can eat many times their weight**: J. E. Thorpe, C. Talbot, M. S. Miles, C. Rawlings, and D. S. Keay, "Food Consumption in 24 Hours by Atlantic Salmon (*Salmo salar* L.) in a Sea Cage," *Aquaculture* 90, no. 1 (1990): 41–47. See also Clare Leschin-Hoar, "90 Percent of Fish We Use for Fishmeal Could Be Used to Feed Humans Instead," NPR, Feb. 13, 2017, https://tinyurl.com/y8yzturq.

157 **Up to 70 percent of the food**: Xinxin Wang, Lasse Mork Olsen, Kjell Inge Reitan, and Yngvar Olsen, "Discharge of Nutrient Wastes from Salmon Farms: Environmental Effects, and Potential for Integrated Multi-trophic Aquaculture," *Aquaculture Environment Interactions* 2 (2012): 267–283.

158 **they lose their genetic instinct**: C. Roberge et al., "Genetic Consequences of Interbreeding Between Farmed and Wild Atlantic Salmon," *Molecular Ecology* 17, no. 1 (Jan. 2008): 314–324. See also Rebecca Clarren, "Genetic Engineering Turns Salmon into Fast Food," *High Country News*, June 23, 2003, https://tinyurl.com/y7fscnly.

158 **a disease known as ISA**: M. Aldrin et al., "Modeling the Spread of Infectious Salmon Anaemia Among Salmon Farms Based on Seaway Distances Between Farms and Genetic Relationships Between Infectious Salmon Anaemia Virus Isolates," *Journal of the Royal Society Interface* 8, no. 62 (2011): 1346–1356.

158 **salmon farms spread SAV**: M. D. Jansen et al., "The Epidemiology of Pancreas Disease in Salmonid Aquaculture: A Summary of the Current State of Knowledge," *Journal of Fish Diseases* 40, no. 1 (May 2016): 141–155.

158 **sea lice kill about 50,000 wild salmon a year**: Stephen Castle, "As Wild Salmon Decline, Norway Pressures Its Giant Fish Farms," *New York Times*, Nov. 6, 2017.

160 **a chemical marketed as Slice**: You Song et al., "Whole-Organism Transcriptomic Analysis Provides Mechanistic Insight into the Acute Toxicity of Emamectin Benzoate," *Environmental Science and Technology* 50 (2016): 11994–12003. See also Norwegian Institute for Water Research, "Anti-Sea Lice Drugs May Pose Hazard to Non-target Crustaceans," Phys.org (Jan. 20, 2017); and John Vidal, "Salmon Farming in Crisis," *Guardian*, Apr. 1, 2017, https://tinyurl.com/m9964xt.

161 **These cages, called Eggs**: Aarskog sees closed containment aquaculture as a key to the so-called Blue Revolution. See Trond W. Rosten, "New Approaches to Closed-Containment at Marine Harvest," Feb. 9, 2015, https://tinyurl.com/y93stf2l. See also Emiko Terazono, "Norway Turns to Radical Salmon Farming Methods," *Financial Times*, Mar. 13, 2017.

161 **gained support from environmental groups**: Personal communication, Ingrid Lomelde, Apr. 2018.

162 **The Chinese first began farming carp around 1000 BC:** The FAO offers an accessible history of fish farming in Herminio R. Rabanal, "History of Aquaculture," 1988, at https://tinyurl.com/zzjyv9r.

163 **serious problem of inbreeding:** Ed Yong, "The Scary Thing About a Virus That Kills Farmed Fish," *The Atlantic,* Apr. 5, 2016.

166 **were among the first living organisms:** Russell Leonard Chapman, "Algae: The World's Most Important 'Plants'—An Introduction," *Mitigation and Adaptation Strategies for Global Change* 18 (Sept. 1, 2010): 5–12. See also Nick Stockton, "Fattened, Genetically Engineered Algae Might Fuel the Future," *Wired,* June 19, 2017.

167 **it makes no difference to a salmon:** Claes Jonermark, operations director for the fish feed division at Marine Harvest, patiently explained the science of fishfeed. See also Timothy B. Wheeler, "Repairing Aquaculture's Achilles' Heel," *Baltimore Sun,* Aug. 19, 2012; and International Union for Conservation of Nature, "Vegetarian Feed One of the Keys to Sustainable Fish Farming," June 12, 2017, https://tinyurl.com/y9vlhzfs.

167 **five pounds of wild fish:** "Fish In: Fish Out (FIFO) Ratios for the Conversion of Wild Feed to Farmed Fish, Including Salmon," IFFO—The Marine Ingredients Organisation, 2015, https://tinyurl.com/yby6epzx.

167 **0.7 pounds of wild fish for every pound of salmon:** Personal communication, Aarskog.

168 **for nearly half of humanity:** "Many of the World's Poorest People Depend on Fish," FAO, June 7, 2005, https://tinyurl.com/ybx287x3. See also Marjo Vierros and Charlotte De Fontaubert, *The Potential of the Blue Economy: Increasing Long-Term Benefits of the Sustainable Use of Marine Resources for Small Island Developing States and Coastal Least Developed Countries* (Washington, D.C.: World Bank, 2017).

CHAPTER 8: MEAT HOOKED

170 **nearly doubled worldwide in three decades:** "What the World Eats," *National Geographic,* https://tinyurl.com/yc2d48gg.

170 **expected to double again by 2050:** "Meat and Meat Products," Animal Production and Health, FAO, Apr. 2016, https://tinyurl.com/42w47gz.

170 **Livestock production accounts for about 15 percent:** "Tackling Climate Change Through Livestock," FAO, 2013, https://tinyurl.com/ybsbf8vj.

170 **American adult eats about 100 grams:** Sophie Egan, "How Much Protein Do We Need?" *New York Times,* July 28, 2017, https://tinyurl.com/yasspl3c.

170 **every one was befouled:** Andrea Rock, "How Safe Is Your Ground Beef?" *Consumer Reports,* Dec. 21, 2015, https://tinyurl.com/ybsota6f.

170 **"come in two flavors of unappealing":** Rowan Jacobsen, "This Top-Secret Food Will Change the Way You Eat," *Outside,* Dec. 26, 2014, https://tinyurl.com/y9mbosve.

172 **cloning more than 100,000 head:** Charlie Sorrel, "China Is Building a Million-Embryo-a-Year Cloned Meat Factory," *Fast Company,* Dec. 8, 2015, https://tinyurl.com/ya92vjs8.

172 **an academic researcher who in 2016:** "WTAMU Research Results Signal Potential for Increased Efficiency in Beef Industry" (press release), West Texas A&M University—WTAMU News, June 29, 2016, https://tinyurl.com/zqae663.

172 **the total feed and water required:** Personal communication, Ty Lawrence, Oct. 2016.

173 **weighs about 1000 pounds and produces only about half:** Food Safety Division—Meat Inspection Services, "How Much Meat?" Oklahoma Department of Agriculture, Food, and Forestry, https://tinyurl.com/y9f3mrbk.

174 **reduce the greenhouse gas emissions:** Personal communication, Uma Valeti and Eric Schulze, Oct. 2018.

174 **eliminate the risk of bacterial contamination:** Ibid.

174 **reduce the risk of heart disease and obesity:** Marta Zaraska, "Is Lab-Grown Meat Good for Us?" *The Atlantic,* Aug. 19, 2013, https://tinyurl.com/yapstflr.

174 **Tyson Foods announced plans:** Chloe Sorvino, "Tyson Invests in Lab-Grown Protein Startup Memphis Meats, Joining Bill Gates and Richard Branson," *Forbes,* Jan. 29, 2018, https://tinyurl.com/y8t2w4vy.

174 **one of every five pounds:** Tysonfoods.com, 2018, https://tinyurl.com/yar5xb4z.

174 **Tyson sells $15 billion worth of beef:** Jonathan Poland, "Tyson Foods: A Long-Term Buy Under $65," Yahoo! Finance, Aug. 20, 2018, https://tinyurl.com/yc4gg84v.

174 **$8 billion in prepared foods:** Keith Nunes, "Tyson Foods Showcases Prepared Foods Innovation Tiers at CAGNY," *Food Business News,* Feb. 22, 2018, https://tinyurl.com/ycbh3hn8.

174 **processes about 1.8 billion animals a year:** Tysonfoods.com, 2018, https://tinyurl.com/yar5xb4z.

174 **responsible for more greenhouse gases:** Tyson did not directly disclose their greenhouse gas emissions. In the company's sustainability report, it says, "Approximately 90 percent of our emissions come from our supply chain and are not owned by the company." Their scope 1 and 2 emissions (what they do directly control) amount to about 5.6 million metric tons. When you add in the other 90 percent, Tyson is responsible for about 56 million metric tons of greenhouse gas emissions a year. See also Tyson Foods, "Reducing Our Carbon Footprint," 2017, https://tinyurl.com/yanxzpp6.

174 **the whole of Ireland:** Ireland emits about 59 million metric tons of greenhouse gases a year. See United Nations, Climate Change Secretariat, "Summary of GHG Emissions for Ireland," 2012, https://tinyurl.com/yageo9ek.

175 **raised more than $350 million:** Taylor Soper, "As Funding Nears $400M, Impossible Foods Targets Meat Eaters with Plant-Based Burger That 'Bleeds,'" *Geek Wire,* Apr. 5, 2018, https://tinyurl.com/yalc2czg.

175 **retail sales of meat alternatives:** "Plant-Based Food Options Are Sprouting Growth for Retailers," Nielsen, June 13, 2018, https://tinyurl.com/ybvte4hh.

175 **many times the growth of meat sales:** "Total Consumer Report," Nielsen, June 2018, https://tinyurl.com/yd5h8hgf.

175 **70 percent of meat eaters:** Sharon Palmer, "Shining the Light on Plant Proteins," *Chicago Health,* Sept. 20, 2018, https://tinyurl.com/ybcgv6tz.

175 **"another way to harvest meat":** Jacob Bunge, "Cargill Invests in Startup That Grows 'Clean Meat' from Cells," *Wall Street Journal,* Aug. 23, 2017.

176 **"More than 90 percent of the global population":** Personal communication, Uma Valeti, Oct. 2018. A 2010 study puts the figure at closer to 80 percent: see Eimar Leahy, Sean Lyons, and Richard S. J. Tol, "An Estimate of the Number of Vegetarians in the World," *Economic and Social Research Institute,* Mar. 2010, https://tinyurl.com/y8synv5v.

176 **China, where meat demand is surging:** Marcello Rossi, "Will China's Growing Appetite for Meat Undermine Its Efforts to Fight Climate Change?" *Smithsonian Magazine,* July 30, 2018, https://tinyurl.com/yd2nakc3.

179 **250 billion pounds today to nearly 500 billion:** "How to Feed the World 2050," FAO, Oct. 13, 2009.

179 **"We shall escape the absurdity":** Winston Churchill, "Fifty Years Hence," *Strand Magazine,* Dec. 1931.

181 **a few thousand dollars a pound:** Zara Stone, "The High Cost of Lab-to-Table Meat," *Wired,* Mar. 8, 2018.

184 **cultured bluefin tuna:** Emma Cosgrove, "Finless Foods Raises $3.5m Seed Round to Culture Bluefin Tuna," *Ag Funder News,* June 20, 2018, https://tinyurl.com/y8avowtl.

184 **fetal bovine serum:** Elaine Watson, "SuperMeat Founder: 'The First Company That Gets to Market with Cultured Meat That Is Cost Effective Is Going to Change the World,'" *Food Navigator,* July 20, 2016, https://tinyurl.com/y8nvn72t.

186 **the root nodules of soybeans:** Matt Simon, "The Impossible Burger: Inside the Strange Science of the Fake Meat That 'Bleeds,'" *Wired,* Sept. 20, 2017.

187 **the largest number of U.S. households:** Mahita Gajanan, "The Meat Industry Has Some Serious Beef with Those 'Bleeding' Plant-Based Burgers," *Time,* Mar. 21, 2018.

188 **unexpectedly resigned as CEO:** Tom Polansek, "Tyson Foods CEO Steps Down for Personal Reasons," Reuters, Sept. 17, 2018, https://tinyurl.com/y8c7q0xe.

189 **recalled 130,000 pounds:** Maria Machuca, "Cargill Meat Solutions Recalls Ground Beef Products Due to Possible E. Coli 026 Contamination" (press release), U.S. Department of Agriculture, Sept. 19, 2018, https://tinyurl.com/y74fwh5r.

190 **$200 million market for meat:** The North American Meat Institute says on their website that as of the latest survey in 2003, the American meat industry rakes in roughly $198 billion a year, https://tinyurl.com/y9xtwwh8.

191 **Americans waste more food:** Dana Gunders, "Wasted: How America Is Losing Up to 40 Percent of Its Food from Farm to Fork to Landfill," Natural Resources Defense Council, Aug. 16, 2017, https://tinyurl.com/glxxnt4.

CHAPTER 9: STOP THE ROT

193 **Kroger's stores serve 9 million:** "The Kroger Family of Companies 2018 Sustainability Report," Kroger, 2018, https://tinyurl.com/y8v7tcpr.

193 **75 million pounds:** Ibid.

195 **Fifty-two million tons of food:** "27 Solutions to Food Waste," ReFED, 2018, https://tinyurl.com/y9ar7rha.

195 **25 percent more per capita:** Jonathan Bloom, *American Wasteland: How America Throws Away Nearly Half of Its Food (and What We Can Do About It)* (Cambridge, Mass.: Da Capo Press, 2010). See also U.S. Environmental Protection Agency, "Municipal Solid Waste Generation, Recycling, and Disposal in the United States: Facts and Figures for 2012," Feb. 2014, https://tinyurl.com/y8ec8k6j.

195 **throw out more than a pound of food a day:** Zach Conrad, Meredith T. Niles, Deborah A. Neher, Eric D. Roy, Nicole E. Tichenor, Lisa Jahns, "Relationship Between Food Waste, Diet Quality, and Environmental Sustainability," *PLOS ONE* 13 (2018).

195 **between $162 billion and $218 billion:** "Frequently Asked Questions," U.S. Department of Agriculture, 2010, https://tinyurl.com/y82prs5o. See also Jonathan Bloom, "A New Roadmap for Fighting Food Waste," *National Geographic,* Mar. 14, 2016, https://tinyurl.com/y9uu6wp7.

196 **21 percent of all freshwater:** "27 Solutions to Food Waste," ReFED.

197 **Kroger was the third-highest performer:** Jennifer Molidor and Jordan Figueiredo, "Checked Out: How U.S. Supermarkets Fail to Make the Grade in Reducing Food Waste," Center for Biological Diversity, Apr. 2018, https://tinyurl.com/yc7d8ut8.

197 **resisted efforts to address the food-waste problem:** Heather Haddon and Laura Stevens, "Investors Want to Talk Food Waste with Amazon," *Wall Street Journal,* Mar. 1, 2018.

199 **tossed out significantly more food:** Darby Hoover, "Estimating Quantities and Types of Food Waste at the City Level," National Resources Defense Council, 2017, https://tinyurl.com/y9zm7ax9.

200 **20 percent of all the fruit and vegetables:** JoAnne Berkenkamp, "Beyond Beauty: The Opportunities and Challenges of Cosmetically Imperfect Produce," Minnesota Institute for Sustainable Agriculture, May 2015, https://tinyurl.com/ybncfs6c.

200 **produce flavors and antioxidants when they're under stress:** Jo Robinson, *Eating on the Wild Side: The Missing Link to Optimum Health* (New York: Little, Brown, 2013).

201 **reducing that country's food waste:** Zlata Rodionova, "Denmark Reduces Food Waste by 25% in Five Years with the Help of One Woman—Selina Juul," *Independent,* Feb. 28, 2017.

201 **reducing banana waste by 90 percent:** Ibid.

202 **France recently passed a law:** Eleanor Beardsley, "French Food Waste Law Changing How Grocery Stores Approach Excess Food," NPR, Feb. 24, 2018.

202 **cutting per capita food waste in half:** "EU Actions Against Food Waste," European Commission, https://tinyurl.com/ya36hlhy.

203 **spend a much smaller portion:** Alex Morrell and Andy Kiersz, "Seeing How the Highest and Lowest-Earners Spend Their Money Will Make You Think Differently About 'Rich' vs. 'Poor,'" *Business Insider,* Dec. 4, 2017, https://tinyurl.com/y75rt906.

204 **gene edited to eliminate browning:** "How'd We 'Make' a Nonbrowning Apple?" Arctic Apples, https://tinyurl.com/ybgpqz83.

206 **Food Recovery Act legislation:** Food Recovery Act of 2017, S. 1680, 115th Congress (2017–2018).

208 **90 million pounds of food a year:** Personal communication, Nick Cortolillo, plant manager, K.B. Specialty Foods, Feb. 2018.

209 **The first anaerobic digester was built in India:** Tasneem Abbasi, S. M. Tauseef, and S. A. Abbasi, "A Brief History of Anaerobic Digestion and 'Biogas,'" SpringerBriefs in Environmental Science, vol. 2, *Biogas Energy* (Springer, 2012).

210 **Since 2007, every household:** "Zero Waste Case Study: San Francisco," Environmental Protection Agency, https://tinyurl.com/y9tpm50p.

210 **23 million tons of food and yard waste:** "National Overview: Facts and Figures on Materials, Wastes and Recycling," Environmental Protection Agency, https://tinyurl.com/ycn7td8f.

211 **"Zero Waste City" by 2030:** "Draft: Livable Nashville," Office of Mayor Megan Barry, https://tinyurl.com/ybarokvg.

211 **70 percent of the world's freshwater:** "Water for Sustainable Food and Agriculture," FAO, 2017, https://tinyurl.com/yaskcsre.

CHAPTER 10: PIPE DREAMS

213 **less rainfall than any other period in at least nine hundred years:** "NASA Finds Drought in Eastern Mediterranean Worst of Past 900 Years," NASA, Mar. 1, 2016, https://tinyurl.com/j48jxbg.

213 **95 percent agriculturally self-sufficient:** Corinne Sauer and Shael Kirshenbaum, "Israelis Give More than NIS 4 Billion a Year in Subsidies to Farmers," *Jerusalem Post,* Jan. 1, 2014, https://tinyurl.com/yaut3e5g.

213 **It takes a gallon of water:** M. M. Mekonnen and A. Y. Hoekstra, "The Green, Blue and Grey Water Footprint of Crops and Derived Crop Products," *Hydrology and Earth System Sciences* 15 (2011): 1577–1600.

213 **about 80 percent of Israel's water supply:** Personal communication, Amir Peleg, Aug. 2018.

213 **about 21 billion gallons per year:** Isabel Kershner, "A Rare Middle East Agreement, on Water," *New York Times,* Dec. 9, 2013.

213 **residents of the West Bank receive less than half the freshwater supply:** "Israel's Water Challenge," *Forbes,* Dec. 26, 2013, https://tinyurl.com/ycueyhlc.

214 **About a quarter of Palestinians are food-insecure:** "Socio-Economic & Food Security Survey 2014: State of Palestine—Report," United Nations, May 2016, https://tinyurl.com/ya7bzn24.

216 **utilities worldwide lose about a third of the water:** Bill Kingdom, Roland Liemberger, and Philippe Marin, "The Challenge of Reducing Non-Revenue Water (NRW) in Developing Countries," World Bank, Dec. 2006, https://tinyurl.com/y9e94mop. See also Alan Wyatt, Jennifer Richkus, and Jemima Sy, "Using Performance-Based Contracts to Reduce Non-Revenue Water," International Bank for Reconstruction and Development/The World Bank, June 2016.

216 **leakage is about 60 percent:** Personal communication, Avshalom Felber, June 2015.

216 **waste up to about 30 percent of the water:** Heather Clancy, "With Annual Losses Estimated at $14 Billion, It's Time to Get Smarter About Water," *Forbes,* Sept. 19, 2013, https://tinyurl.com/y7gj6svw.

217 **70 percent desertified:** "First National Report on the Implementation of United Nations Convention to Combat Desertification," The Blaustein

Institute for Desert Research, Nov. 2000, https://tinyurl.com/ycjj3ekk. See also Alon Tal, "To Make a Desert Bloom: The Israeli Agriculture Adventure and the Quest for Sustainability," *Agricultural History* 82, no. 2 (Spring 2007): 228–257; and Jon Felder, "Focus on Israel: Israel's Agriculture in the 21st Century," Israel Ministry of Foreign Affairs, Dec. 24, 2002.

217 **The drip irrigation was more than twice as efficient:** David Shamah, "What Israeli Drips Did for the World," *Times of Israel*, Aug. 20, 2013.

218 **a commercial process for removing salt from seawater:** Rivka Borochov, "Israel Leads Way in Making Saltwater Potable," Israel Ministry of Foreign Affairs, Jan. 31, 2012, https://tinyurl.com/y9qso2p9.

218 **Israel now recycles more than 85 percent of the water supply:** Jewish National Fund, "Water Solutions: Solutions for a Water-Starved World," https://tinyurl.com/ybsgmz3h.

219 **Hagihon loses about 10 percent of its water:** "Israel's Largest and Most Advanced Regional Water and Wastewater Utility," Hagihon Company Ltd., https://tinyurl.com/ycv8rc4d.

220 **20 percent of utilities globally:** Personal communication, Amir Peleg.

220 **ten dollars per thousand gallons:** Sarah Frostenson, "Water Is Getting Much, Much More Expensive in These 30 Cities," *Vox*, May 19, 2017.

220 **in a state of severe drought:** Michael Dettinger, Bradley Udall, and Aris Georgakakos, "Western Water and Climate Change," *Ecological Applications* 25, no. 8 (2015).

221 **Egypt will approach a state of "absolute water crisis":** Mohamed Ezz and Nada Arafat, "'We Woke Up in a Desert'—The Water Crisis Taking Hold Across Egypt," *Guardian*, Aug. 4, 2015.

221 **its water demand is expected to double:** Lara Nassar and Reem Al-Haddadin, "A Guidance Note for SDG Implementation in Jordan," West Asia–North Africa Institute, Nov. 2017, https://tinyurl.com/yb7y69y8.

221 **over half of the population will need to be relocated:** Somini Sengupta, "Warming, Water Crisis, Then Unrest: How Iran Fits an Alarming Pattern," *New York Times*, Jan. 18, 2018.

221 **imposed a ban on Israeli imports:** "South Africa's Ruling Party Endorses Boycott of Israel," *Times of Israel*, Dec. 21, 2012, https://tinyurl.com/ybhonjdw.

221 **agreed to drop the ban:** Luke Tress, "As 'Day Zero' Looms, South Africa Open to Israeli Water Tech, Researcher Says," *Times of Israel*, Mar. 8, 2018, https://tinyurl.com/y8a4hl2d.

221 **signed an agreement for a blue-tech transfer:** "California and Israel Sign Pact to Strengthen Economic, Research Ties," Office of Gov. Edmund J. Brown Jr., Mar. 5, 2014, https://tinyurl.com/y9dgjnz9.

221 **a pipe burst in Los Angeles:** Joseph Serna, James Queally, Larry Gordon, and Caitlin Owens, "Broken Water Main Floods UCLA at 75,000 Gallons per Minute," *Los Angeles Times,* July 30, 2014.

222 **California recycles about 15 percent of its wastewater:** Heather Cooley, Peter Gleick, and Robert Wilkinson, "Water Reuse Potential in California," Pacific Institute, June 2014, https://tinyurl.com/y8u4m92w.

222 **climbing nearly 10 percent a year:** Alastair Bland, "Californians Are Struggling to Pay for Rising Water Rates," KQED, Feb. 27, 2018, https://tinyurl.com/y7sdb9jk.

223 **"If we could ever competitively":** Quoted in Jeff Hull, "Water Desalination: The Answer to the World's Thirst?" *Fast Company,* Feb. 1, 2009, https://tinyurl.com/y778gdpv.

225 **100 million gallons of drinking water:** Greg Mellen, "From Waste to Taste: Orange County Sets Guinness Record for Recycled Water," *Orange County Register,* Feb. 18, 2018, https://tinyurl.com/yddubtdu.

226 **35 percent of its water from recycled sewage by 2030:** Joshua Emerson Smith, "San Diego Will Recycle Sewage into Drinking Water, Mayor Declares," *San Diego Tribune,* May 10, 2017, https://tinyurl.com/lu9v8yr.

CHAPTER 11: DESPERATE MEASURES

229 **"This was the shot that pierced":** Verna Aardema, *Bringing the Rain to Kapiti Plain* (New York: Dial Books for Young Readers, 1981).

230 **More than 80 percent:** Personal communication, Eknath Khadse, July 2015.

230 **scientists attributed to climate change:** Dana Nuccitelli, "2013 Was the Second-Hottest Year Without an El Niño Since Before 1850," *Guardian,* Feb. 6, 2014.

230 **dropped by more than a third:** Personal communication, Eknath Khadse.

230 **farmers in Maharashtra had committed suicide:** Zeeshan Shaikh, "Maharashtra Tops 2015 Suicide Chart," *Indian Express,* Jan. 2, 2017, https://tinyurl.com/ycsjbzmt.

232 **practiced in the United States since the 1940s:** Virginia Simms, "Making the Rain: Cloud Seeding, the Imminent Freshwater Crisis, and International Law," *The International Lawyer* 44 (2010): 915–937.

232 **China spends hundreds of millions of dollars a year:** George Dvorsky, "China's Ambitious New Rain-Making System Would Be as Big as Alaska," Gizmodo, Apr. 25, 2018, https://tinyurl.com/y8cx8hmw.

232 **nontoxic and has shown negligible impact:** D. M. Ramakrishna and T. Viraraghavan, "Environmental Impact of Chemical Deicers—A Review," *Water, Air, and Soil Pollution* 166 (2005): 52–56.

233 **increase in precipitation:** Sophie Quinton, "'Cloud Seeding' May Make It Snow, but It Will Reduce Droughts in the West," *Washington Post*, Feb. 26, 2018, https://tinyurl.com/ybls2wb2.

234 **the government gives what's known as a "compensation":** Gopala Sarma Poduri, "Short-Term Cost of Suicides in India," *Indian Journal of Psychological Medicine* 38 (2016): 524–528.

234 **the worst water crisis in its history:** "Composite Water," NITI Aayog, June 2018.

234 **200,000 Indians are dying annually:** Annie Banerji, "India's 'Worst Water Crisis in History' Leaves Millions Thirsty," Reuters, July 4, 2018.

234 **drought intensity has increased 15 to 20 percent:** National Oceanic and Atmospheric Administration data on drought patterns over time can be found at https://tinyurl.com/y7ju84bs.

235 **farm-swallowing floods:** Lynh Bui and Breena Kerr, "Repeated Natural Disasters Pummel Hawaii's Farms," *Washington Post*, Aug. 28, 2018.

235 **four million Ethiopians were receiving emergency food rations:** "UN Says 4.5 Million Ethiopians Now in Need of Food Aid After Poor Rains," *Guardian*, Aug. 25, 2015.

236 **people needing emergency food had doubled:** "Ethiopia Appeals for International Aid in Face of Deepening Food Insecurity," *Guardian*, Oct. 15, 2015.

236 **10.2 million people needed food:** "Ethiopia: 10.2 Million Ethiopians Need Emergency Food Assistance," USDA Foreign Agricultural Service, GAIN Report Number ET1543, Dec. 28, 2015, https://tinyurl.com/y8jy9nye.

236 **feed more than 18 million people:** Aislinn Laing, "Ethiopia Struggles with Worst Drought for 50 Years Leaving 18 Million People in Need of Aid," *Telegraph*, Apr. 23, 2016.

236 **almost $800 million across eighteen months:** Personal communication, Mitiku Kassa, Dec. 2016.

236 **aid partners contributed another $700 million:** Ibid.

238 **ninety different languages:** M. L. Bender, "The Languages of Ethiopia: A New Lexicostatistic Classification and Some Problems of Diffusion," *Anthropological Linguistics* 13 (1971): 170.

239 **Civilian protests in 2017:** "Ethiopia: Events of 2017," Human Rights Watch, 2018, https://tinyurl.com/ydg3ym4g.

239 **fastest-growing economies in Africa:** Chris Giles, "Ethiopia Is Now Africa's Fastest Growing Economy," CNN, Apr. 24, 2018.

239 **agricultural productivity has doubled:** "Country Indicators: Ethiopia," FAO database, https://tinyurl.com/y9z52bdh.

239 **12 billion gallons of drinking water:** Personal communication, John Aylieff, Dec. 2016.

240 **"no notable increase in infant or child mortality rates":** The UNICEF data on Ethiopia's infant and child mortality rates over time can be found at https://tinyurl.com/yd2xf7qc.

240 **twenty times higher than seed provisions:** Personal communication, Alemu Manni, Aug. 2018.

241 **yields increased 20 percent:** Ibid.

CHAPTER 12: ANTIQUITY NOW

244 **More than two billion people:** "Measuring the Earth's Dry Forests," NASA Earth Observatory, 2015, https://tinyurl.com/y7g70wto.

244 **it produces edible leaves:** Maanvi Singh, "My Mom Cooked Moringa Before It Was a Superfood," NPR, Sept. 21, 2015.

244 **high in omega-3 fatty acids:** Ramesh Kumar Saini, Iyyakkannu Sivanesan, and Young-Soo Keum, "Phytochemicals of *Moringa oleifera*: A Review of Their Nutritional, Therapeutic and Industrial Significance," *3 Biotech* 6 (2016): 203.

244 **strong anti-inflammatory:** Jeff Fahey, "*Moringa oleifera:* A Review of the Medical Evidence for Its Nutritional, Therapeutic, and Prophylactic Properties. Part 1," *Trees for Life Journal,* Dec. 1, 2005, https://tinyurl.com/ya8jersd.

244 **antidiabetic properties:** R. Gupta, M. Mathur, V. K. Bajaj, P. Katariya, S. Yadav, R. Kamal, and R. S. Gupta, "Evaluation of Antidiabetic and Antioxidant Activity of *Moringa oleifera* in Experimental Diabetes," *Journal of Diabetes* 4 (2012): 164.

252 **five times more omega-3s than modern varieties:** Robinson, *Eating on the Wild Side,* 190.

253 **seven times more phytonutrients:** Jo Robinson, "Breeding the Nutrition Out of Our Food," *New York Times,* May 25, 2013.

253 **the "junk food effect":** Helena Bottemiller Evich, "The Great Nutrient Collapse," The Agenda, *Politico,* Sept. 13, 2017, https://tinyurl.com/yadrz3n9.

253 **declined 3 to 17 percent:** Personal communication, Samuel Myers, Oct. 2018. See also Danielle E. Medek, Joel Schwartz, and Samuel S. Myers, "Estimated Effects of Future Atmospheric CO_2 Concentrations on Protein Intake and the Risk of Protein Deficiency by Country and Region," *Environmental Health Perspectives* 125 (2017).

254 **one in ten Americans:** "State Indicator Report on Fruits and Vegetables," Centers for Disease Control and Prevention (CDC), 2018, https://tinyurl.com/ybpc6n42.

254 **65 percent of the U.S. population:** "Fruit and Vegetable Consumption Among Adults—United States, 2005," CDC, 2007, https://tinyurl.com/y92txlwu.

254 **Nearly 25 million Americans:** Angela Hilmers, David C. Hilmers, and Jayna Dave, "Neighborhood Disparities in Access to Healthy Foods and Their Effects on Environmental Justice," *American Journal of Public Health* 102 (2012): 1644–1654.

254 **jumped more than 30 percent:** "U.S. Adult Consumption of Added Sugars Increased by More than 30% over Three Decades," *ScienceDaily*, Nov. 4, 2014, https://tinyurl.com/yc86sk09.

255 **increased about 20 percent:** C. D. Fryar, Q. Gu, C. L. Ogden, and K. M. Flegal, "Anthropometric Reference Data for Children and Adults: United States, 2011–2014," National Center for Health Statistics, *Vital and Health Statistics* 3, no. 39 (2016), https://tinyurl.com/yaupjq05.

255 **increased 700 percent:** "Long-Term Trends in Diabetes," CDC, Apr. 2017, https://tinyurl.com/yartp20q.

257 **sequenced the genome in 2017:** David E. Jarvis et al., "The Genome of *Chenopodium quinoa*," *Nature* 542 (2017): 307–312.

257 **developed GMO cassava:** Adewale Oparinde, Tahirou Abdoulaye, Djana Babatima Mignouna, and Adebayo Simeon Bamire, "Will Farmers Intend to Cultivate Provitamin A Genetically Modified (GM) Cassava in Nigeria? Evidence from a *k*-Means Segmentation Analysis of Beliefs and Attitudes," *PLOS ONE* 12, no. 7 (2017).

257 **engineered pearl millet:** Pallava Bagla, "'Magic Millet' Gets an Enrichment Boost to Cure Anaemia," *Economic Times*, June 14, 2015, https://tinyurl.com/y7fpc3pk.

257 **zinc-enriched rice:** Elemarie Lamigo-Rosellon, "Improving Health and Nutrition Through Rice Science," July 31, 2017, posted on HarvestPlus.org, https://tinyurl.com/y9xewwfl.

258 **"super bananas":** Melissa Hellmann, "Researchers Hope 'Super Bananas' Will Combat Vitamin A Deficiency," *Time*, June 16, 2014. See also Alon Mwesigwa, "Can a GM Banana Solve Uganda's Hunger Crisis?" *Guardian*, Dec. 12, 2017.

CHAPTER 13: WHAT ROUGH FEAST

261 **deposit liquid polymers:** Samuel Clark Ligon, Robert Liska, Jürgen Stampfl, Matthias Gurr, and Rolf Mülhaupt, "Polymers for 3D Printing and Customized Additive Manufacturing," *Chemical Reviews* 117 (2017): 10212–10290.

265 **a "three-year bread" recipe:** Donna Miles, "Military Explores New Processes, Packaging for Combat Rations," U.S. Department of Defense, *DoD News*, Sept. 28, 2012, https://tinyurl.com/y9bdvqpo.

265 **oxygen-scavenging plastics:** Simon Angelo Cichello, "Oxygen Absorbers in Food Preservation: A Review," *Journal of Food Science and Technology* 52 (2015): 1889–1895.

265 **"three-year pizza":** Emma Graham-Harrison, "The Eat of Battle—How the World's Armies Get Fed," *Guardian*, Feb. 18, 2014.

267 **invested in start-ups:** John Kell, "Campbell Soup Invests in Nutrition Tech Startup," *Fortune*, Oct. 26, 2016, https://tinyurl.com/zw4z8pz; and John Kell, "Nestle Leads $77M Funding Round in Meal-Delivery Startup Freshly," *Fortune*, June 20, 2017, https://tinyurl.com/y9rwrtmd.

267 **the maximum age allowed is thirty-nine:** "Age Requirement by Service," Today's Military, 2018, https://tinyurl.com/ycm9jr3f.

267 **more Latino enlistees than ever before:** Kim Parker, Anthony Cilluffo, and Renee Stepler, "6 Facts About the U.S. Military and Its Changing Demographics," Pew Research Center, Apr. 13, 2017, https://tinyurl.com/ycnamg3p.

268 **Gen Z now accounts for:** L. J. Miller and Wei Lu, "Gen Z Is Set to Outnumber Millennials Within a Year," *Bloomberg*, Aug. 20, 2018, https://tinyurl.com/y7qqspde.

271 **radiant zone drying:** Moumita Chakraborty, Mark Savarese, Eileen Harbertson, James Harbertson, and Kerry L. Ringer, "Effect of the Novel Radiant Zone Drying Method on Anthocyanins and Phenolics of Three Blueberry Liquids," *Journal of Agricultural and Food Chemistry* 58 (2010): 324–330.

271 **Microwave vacuum drying:** G. Ahrens, H. Kriszio, and G. Langer, "Microwave Vacuum Drying in the Food Processing Industry," in *Advances in Microwave and Radio Frequency Processing* (New York: Springer, 2006), 426–435.

273 **total recommended daily nutrition:** "Dietary Guidelines for Americans 2015–2020," U.S. Department of Agriculture, 2015, https://tinyurl.com/ycfkhdwq.

Acknowledgments

IN OUR HOUSE, as in many others, family dinners begin with a round of gratefuls. Gratitude is to food as butter is to bread and curry is to rice—a source of depth and flavor. I give sincere and humble thanks to the legions of food growers and producers and preparers the world over who make it possible for so many of us to enjoy the diversity of flavors and nutrients we have access to every day. For all its flaws and inequities, the sprawling global food system is a full-on, flat-out marvel.

I owe the greatest measure of gratitude to my husband, Carter, who is my primary source of nourishment—intellectual, emotional, and comedic. This book would have been far more difficult, and perhaps impossible, to write without his love and levity. Our kids, Aria and Nicholas, were heroically patient and high-spirited throughout this long process. Instead of complaining about my nights and weekends on the road and at the keyboard, they often tagged along or set up shop in my office and kept me company. This book is theirs as much as it is mine.

I value the wisdom and friendship of my agent, Kimberly Witherspoon, beyond measure. She is far more involved in the food world than I am, and had she not encouraged this project when it was just an idea, I never would have pursued it. Thank you to Heather Jackson for championing this book at the outset, and to my tenacious editor, Diana Baroni, for seeing it through. I'm indebted to Diana for her candor, drive, patience, and great ear for narrative momentum. I feel very lucky to have been taken under her wing. To Michele Eniclerico and the design, production, and marketing teams at Harmony Books, thank you for your excellence and rigor.

Brad Wieners is enormously kind and prodigiously talented. He helped me structure and polish these pages, and in doing so proved to be the world's best caffeinated machete-wielding word ninja. Lindsey Rome agreed to help me cull and curate the photos in this book even though her talent far exceeds a project like this; I so appreciate her generosity of time and spirit.

A book is only as good as its sources, so I'm immeasurably grateful to the central figures in each chapter who invited me into their work and lives, and to the scientists, engineers, and thinkers who shared with me far more time and many more insights than are reflected in this book. They include Michael Pollan, Tamar Haspell, David Lobell, Paul Hawken, Jeanne Nolan, David Friedberg, Jerry Hatfield, Ruth DeFries, Nathanael Johnson, Danielle Nierenberg, Pamela Ronald, Darby Hoover, Gil Gullickson, Amaya Atucha, Alemu Mani, Marcia Ishii-Eitman, Kendra Klein, Dana Perls, Martin Bloem, Sam Myers, Daniel Mason D'Croz, Willy Foote, Adam Davis, Chris and Annie Newman, Willy Pell, Nathan Reed, Ola Helge Hietland, Josh Goldman, Eric Newman, Findle Zhao, Caleb Harper, Sam Kennedy, Ty Lawrence, Gus Vanderberg, Celeste Holz-Schlesinger, Pete Pearson, Emily Broad Leib, Snehal Desai, Lisa Curtis, Rose Wang, John Robinson, Brian Heimberg, Mark Lynas and Manuela Zoninsein.

Becca Richardson provided years of invaluable research and encouragement; she's huge-hearted, hard-working, wildly talented, and capable of everything. I also offer thanks and raise my hat to the redoubtable Ania Szczesniewski, who patiently ferreted out many a fact and, as a farmer and food-justice activist, continually challenged me to consider the ethics of food production. My former student Caroline Saunders did the same and provided valuable research into food history.

I'm grateful to Clark Williams Derry for his inspired wordsmithery and to Cindy Kershner, who lent her meticulous eye to proofing these pages, along with the use of Sewannee Aerie as a writing refuge. I'm also deeply grateful to the Mesa Refuge for offering me a fellowship to write in that oasis of quiet and calm; to Greta Gaines and her family for sharing their Alabama cabin; and to Kathryn Schulz and Casey Cep for sharing their double-wide, baked goods, and incomparable example of a writing life fully lived.

I want to acknowledge other writer friends and mentors who keep me loyal to this profession by continually pushing the bounds of what it can do, in particular, my alma gemela, Alice Randall, Juliet Eilperin, Rebecca Paley, Alix Barzelay, Florence Williams, Jon Meacham, Nick Thompson, Amely Greeven, Caroline Williams, Miranda Purves, and Ben Austen. I received valuable support and motivation along the way from my students and colleagues at Vanderbilt, including Dana Nelson, who provided generous research funds, Cecilia Tichi, Teresa Goddu, Tiffiny Tung, and Steve Werneke.

My mom, Nancy, offered indefatigable support in her cooking, grandparenting, relentless curious mind, and modeling of a passionate career. I'm grateful to my father, Rufus, for his love of storytelling, history, and enterprise, and his devotion to land well tended. My stepfather, Colden Florance, and my stepmother, Hope Griscom, are unreasonably supportive and patient. My mother-in-law, Lyn Little, has invested heart and soul in my family and work, and sent countless thoughtful article clippings (for which the U.S. Postal Service is also appreciative). My brothers, Rufus and Bronson Griscom, are my best friends and trusted advisors on this and all else I do. I'm grateful to Sophie Simmons and Courtney Little, who fed me and my kids more times than I can count, and to Ali Kominsky, Sarah Troy Clark, Evie Kennedy, Torrey Morgan, Daniela Falcone, and Lisa Muloma, who were among the village of people who helped care for my kids when I couldn't. Christina Mangurian, Sarah Douglas, Lissa Smith, Tanaz Eshaghian, Alex Kerry, Keith Meacham, Vandana Abramson, Pauline Diaz, and the Shakti community on Music Row are strongholds of kindness and sanity.

Lastly, I am grateful beyond measure for the memory and extraordinary example of Olivia Taylor Barker, 1974–2014. I didn't try my hand at journalism until she gave me an assignment at the Brown *Independent* our sophomore year in college, and she's been at the heart of every story I've read and told since. No one has loved a good yarn or the pleasures of great food more and better than Olivia. She left this world too soon but remains mercifully present.

Index

A
Aardema, Verna, 229–30
Aarskog, Alf-Helge, 149–55, *154*, 158–62, 164–65, 166–68
Abbey, Edward, 129
Abundant Robotics, 55
"accessibility gap," 27–28, 196, 284
acidification, 157
Adamu, Hassan, 78
Adelman, Jessica, 197
Adesina, Akinwumi, 68
aerobic digestion, 210
AeroFarms Systems, 135–47, *145*, 253–54
aeroponics, 130–31, 138, 140–42. *See also* indoor food production
aesthetics, and food waste, 199–200
Africa
  Bt-maize and Liyago, 79–83
  cacao farms, 50
  drought in, 58, 61–62, 64–67, 71, 80, 83–84, 220–21, 235–42
  global population, 147
  GMO seeds and Monsanto, 61–73, 78–83
  Kassa and famine relief, 235–42
  Matete farm, 84–86
  Oniang'o and ROP, 57–66, 78–79, 88–89
  Shiyuka farm, 61–64
  WEMA program, 65–67, 71–73, 79
Africa Food Prize, 68
African Agricultural Technology Foundation, 59–60, 62
African Centre for Biodiversity, 66, 72
African Development Bank, 68
agave, 84
Agent Orange, 89
agrarianism, 37
agribusiness (industrial agriculture), 19–21, 23
  achievements of, 5
  disadvantages of, 5–6
agricultural imperialism, 72
agricultural runoff, 21, 89–90
agriculture
  history of, 14–17, 18–22
  water use in, 213, 217–18
air quality, in China, 112
Akweneno, Jonas, 86
algae, 162, 165–68, 251
algae blooms, 21, 88, 89
Almería, Spain, greenhouses, 132, *132*
almonds, and water usage, 213
Alzheimer's disease, 103

amaranth, 118–19, 245
Amazon, 127, 140, 197, 276
Aminzadeh, Sara, 224
Ample Meal, 7
anaerobic digestion, 209–11
ancient foods and crops, 243–59
  Kernza, 245, 250–51, 255, 259
  Moringa, 243–45, 246–49, 252, 255–59
  quinoa, 249–50, 255, 257, 258, 259
ancient Greece, 224
ancient Mesopotamia, 14–15, 16
ancient Rome, 51, 130, 131, 200
Andrés, José, 249
Andresen, Jeff, 45–47, 52
anemia, 258
animal cloning, 172–73
animal cruelty, 22–23, 170
Antarctic ice shelves, 44
Anthony, David, 134–35
antibiotics, 268
antioxidants, 35, 188, 200, 252
Apeel Sciences, 207
"apple a day," 35
apples, 30–44
  climate change's impact, 45–47, 54–56
  cultivation, 34–35, 38, 39–44
  Ferguson's farm, 29–42, 52–56
  nutrition, 253
  storage, 35–36
  Wisconsin freeze of 2016, 30–33, 32, 36–37, 40, 42–43, 44–45, 52–53
aquaculture, 148, 149–68
  climate change and fisheries, 155–57, 163–64
  fish feed, 157, 159, 166–68
  Marine Harvest and Aarskog, 149–55, 158–62, 165–68
  problem of fish waste, 155, 161–62, 163, 164–65
  sea lice and, 150–52, 158–61
Aquaculture Stewardship Council, 150
Arab Spring, 18
Arctic apples, 76
Arizona State University, 253
Army Research Development and Engineering Center, Food Innovation Laboratory, 260–61, 264–66, 270–71
Arom, Zvi, 220
artificial intelligence (AI), 25, 105–9, 111, 138, 160
Asparagopsis, 164
aspartame, 24
Aswani, Jane Beth, 86
Atomico, 175
atrazine, 89
Atucha, Amaya, 43, 294*n*
Australia, 100–101, 163, 220, 232
Australis Aquaculture, 155, 162, 163–64
avocados, 50, 207, 213
Aylieff, John, 236, 237, 239–40
Ayurvedic medicine, 169–70
Azam-Ali, Sayed, 248–49

B

baby greens, 131, 134–44
Bacillus subtilis, 71
Bacillus thuringiensis (Bt), 71, 77, 81–83
backyard vegetable garden, 11–13, 22, 146–47, 153, 264
bananas, 64, 199, 201, 258
Barber, Dan, 11
barley, 69, 220, 257
Barnard, Matt, 127
barramundi, 155, 163–64
Barrett, Ann, 270–71
Bayer, 58, 60, 62, 66, 71, 92
Beast Burger, 188
beauty standards, and food waste, 199–200
beef, 168, 169–91
  carbon footprint of, 23, 168, 171–72
  cloning cattle, 172–73

ethics of eating, 22–23, 169–71
Memphis Meats and cultured meat, 173–85, 189–90
in moderation, 170, 190–91
vegan alternatives, 185–88
bees, 49–50
colony collapse disorder, 21, 90, 95
Beijing, 111, 112, 115, 119, 126
Ben-Gurion, David, 217
Berkowitz, Dan, 265
Bertrand, Benoît, 51
beta-carotene, 76
Beyond Meat, 175, 187–88, 251
Bezos, Jeff, 127, 140
Bhopal disaster, 89
big data, 25, 45–47, 106, 108–9, 208, 216
Bill and Melinda Gates Foundation, 60, 62, 257–58
biofortification programs, 258–59
"biomimicry," 97, 182–83
birth abnormalities, 90
Bittman, Mark, 11
black sea bass, 157
Blass, Simcha, 217–18
"blast-frozen," 8
"blue baby syndrome," 90
bluefin tuna, 184
Blue Revolution, 154
Blue River Technology, 92–102, 105–9
Borlaug, Norman, 20, 25, 68, 70
Botany of Desire, The (Pollan), 11
Bourne, Joel, 168
Bowery Farming, 140
Branson, Richard, 175
breast cancer, 24
Breed, Dan, 232–33
Breivik, John, 160
Bringing the Rain to Kapiti Plain (Aardema), 229–30
Brown, Ethan, 187–88
Brown, Gabe, 171
Brown, Jerry, 221
Brown, Patrick, 186–87
Bt-cotton, 77
Bt-maize (corn), 68, 71, 77, 79–83
burdock, 97

C

cacao, 50
Caddyshack (movie), 159
cadmium, 113, 118
Caixin, 111
calcium, 188, 244, 258, 266
California
Heraud and robotics, 90–98, 105–9
salmon species, 156
water resources, 221–28, 232
California Coastkeeper Alliance, 224
California drought of 2015, 48, 49–50, 221
California wildfires of 2018, 5
calorie overload, 3
calories per acre, 70
calories per dollar, 2, 70
camels, 65
Campbell's Soup Company, 267
cancer, 24, 74, 75, 103, 118, 121
cancer prevention, 244, 253
canning, 8
carbon dioxide (CO_2), 83, 104, 138, 157, 250–51, 253–54
carbon sequestration, 104, 250–51
carbon sink, 141–42, 171
Cargill Meats, 45, 175–76, 189
Carlsbad desalination plant, 222–23, *223*, 225
carnivorism, 153, 169–71
ethics of, 22–23, 169–71
Memphis Meats and cultured meat, 173–85, 189–90
carp, 155, 162
Carson, Rachel, 89
Carswell, K. C., 182–83
Cascadian Farms, 251
catfish, 155, 168

cattle cloning, 172–73
cecropia moth, 71
Center for Biological Diversity, 197
Chang, David, 135, 186
Changzhou Foreign Language School, 121
Chaparro, José, 47
ch'arki, 8
chemical wash, 143
Chen, Junshi, 113, 125
cherries, 45–46
Chez Panisse, 173
Chicken-Flavored Pot Pie, 1–5, 8–9, 270
child mortality rates, 240, 258
"chilling units," 42–43
China
  apple supply, 35
  beef producers in, 172
  cloud seeding, 232
  food production, 111–27
  land purchases, 133
  migration in, 112
  organic certification process, 122, 126
  organic farming, 109, 113–21, 122, 123–26
  rice-fish system, 162
China Ministry of Agriculture, 118, 120
China Ministry of Ecology and Environment, 122
China National Center for Food Safety Risk Assessment, 113, 125
Chinook salmon, 156
Chipotle, 50, 77
Chirps Chips, 251–52
chloroquine, 58
chocolate, 50, 265
Churchill, Winston, 179
Church of Jesus Christ of Latter-Day Saints, 4
circular economy, 211
Civil War, 269
Clara Foods, 185
cleaner fish, 159–60
climate change, 38, 47–51, 285–86
  in Africa, 64–67
  agricultural research and, 45–47
  fisheries and, 155–57, 163–64
  food systems and, 6–7, 9
  Green Revolution and, 21–22
  IPCC report, 6, 7, 9
  survival food industry and, 3–5
  Trump and, 44–45
Climate Institute, 50–51
climate volatility, 5, 6, 41, 46, 137, 245
Clinton, Bill, 68
cloning, 34–35, 172–73
closed containment aquaculture, 161–62
cloud seeding, 230–35
cocklebur, 97
cod, 156–57, 168
coffee, 50–51
Colbert, Stephen, 274
Colorado River, 48, 222
Colorado State University, 131
Columbia Business School, 135
Columbia University, 15, 276
composting, 196, 197–98, 199, 209–11
Consumer Reports, 170
"controlled atmosphere storage," 35
Cook, Tim, 111
corn. *See* maize
Cornell University, 52–53, 130, 131, 141
corn syrup, 19, 58, 73
Costco, 196
cotton, 23, 67, 97, 98–102
Crick, Francis, 78
CRISPR, 75, 110–11, 204, 256, 278
crop heaters, 51–52
crop insurance, 48
crop yields, 22, 48, 66, 113
crossbreeding, 68, 70
cross-fertilization, 18–19
"cryoprotectants," 54
cultured meat, 173–85, 189–90

Curtis, Lisa, 247–48, 256

D

Dahl, Roald, 2, 264
Dale, James, 258
dams (damming), *19*, 20
dandelion greens, 253
dandelions, 96–98
Darwin, Charles, 18–19, 70
data science. *See* big data
date-labeling laws, 205–7
da Vinci, Leonardo, 104–5
Davis, Adam, 102
DDT (dichlorodiphenyltrichloroethane), 24, 89
dead zones, 89, 95
deforestation, 170
DeFries, Ruth, 15, 16, 20–22, 276–77, 290*n*
deinvention vs. reinvention, 25–26, 28
Del Monte Foods, 76
Denmark, and food waste, 201–2
Denver, and food waste, 198, 199
Dern, Bruce, 146
Desai, Snehal, 225, 226, 227–28
desalination, 218, 223–25, 226
devil's claw, 97
DFJ Venture Capital, 175
diabetes, 22, 75, 255
Diamond, Jared, 69
dicamba, 97–98
DiCaprio, Leonardo, 163
Dickinson, Emily, 243
Digita, 94
DNA, 75, 78
"domesticate," 15
domus, 15
dosha, 169–70, 267
DowDuPont Inc., 92
Drawdown (Hawken), 9, 138, 144
drip irrigation, 217–18
drones, 108–9
drought
  in Africa, 58, 61–62, 64–67, 71, 80, 83–84, 220–21, 235–42
  Arab Spring and role of, 17–18
  in California, 5, 48, 49–50, 221
  classification, 234–35
  cloud seeding, 230–35
  food scarcity and famine, 235–42
  GMO seeds, 61–73, 78–80, 82–83. *See also* seeds of drought
  IPCC report, 6
  "New Dust Bowl," 48
  Pentagon warning on impacts, 17
  water resources and, 220–21, 223, 226
DroughtGard seeds, 71, 83
DroughtTego seeds, 61–67, 70–72, 85, 296*n*
DroughtTela seeds, 67, 71, 77, 80–81
drought tolerance, 38, 67, 71, 76–78, 80, 82–84
duck, 176–78
Dungeness crab, 157
Dust Bowl, 103

E

Earthbound Farm, 115, 116, 143
Eating on the Wild Side (Robinson), 252–53
Eatsa, 249
ecological farming, 26–29, *27*, 279–87
Edge, Mark, 71–72, 83
edible insects, 251–52
Eggs (salmon cages), 161–62, 164–65
Egypt, 17, 180, 221
electric cars, 153, 175, 190
El Niño, 230
emamectin benzoate, 160–61
Emuleche River, 64
environmentalism, 11–12, 22–29, 89
Environmental Protection Agency (EPA), 89, 102, 210–11
Environmental Working Group, 90
Environment International, 121
Escherichia coli (E. coli), 143, 174, 189

Escobar, Pablo, 172
ethanol, 70
ethical eating, 11–12, 22–23
Ethiopia, 3, 36, 64–65, 235–42
Ethiopian protests of 2017, 239
evolution, 18–19, 34, 69

F

Fahey, Jed, 244, 257
Fairchild, David, 169
fall armyworm, 68
famine, 3, 18, 64–65, 235–42
FarmedHere, 139–40
farmed salmon. *See* salmon farming
farmers' markets, 12
farm size, 37
Federal Emergency Management Agency (FEMA), 4
Felber, Avshalom, 218–19
fenthion, 82
Ferguson, Andy, 29–42, 44, 49, 52–56, 53
fertilizers, 88–89, 103, 106
  agricultural runoff and, 21, 89–90
  in China, 112–13, 122
  compost, 190, 196, 209, 210
  history of, 16, 18, 19
  soil erosion and, 102, 103, 133
fetal bovine serum, 182, 184–85
Fields China, 126
Finless Foods, 184
fish farming. *See* aquaculture
fish feed, 157, 159, 166–68
fish waste, 155, 161–62, 163, 164–65
Fitbit, 266, 268
flavonoids, 274
FlavrSavr tomatoes, 76
floods (flooding), 4, 6, 44, 77, 132–33, 235
"flora-based" proteins, 185
fly-fishing, 153
food additives, 24, 186, 251, 256
food aid, 64–65, 236
Food and Agriculture Organization (FAO), 240–41, 242
Food and Chemical Toxicology, 74
Food and Drug Administration (FDA), 76, 206, 273
food aromas, 187
food banks, 193, 202, 204–5
food coloring, 24
Food Cowboy, 205
Food Date Labeling Act, 206
food deserts, 254
food ethics, 11–12, 22–29
food forests, 282, 285–87
Foodini (robot), 260–63, 277
food insecurity, 65, 76, 196
food inspections, 113
food labels, 193, 195, 205–7
food leftovers, 199, 203
food nostalgia, 12, 114, 116–17, 170
food packaging, 207–8
food preservation, 16, 207
food prices, 2, 5, 7, 23, 28
Food Recovery Act, 206–7
food scarcity, 21, 190, 235–42
food scraps, 208–11
food storage, 16
food surpluses, 17, 21–22
food systems
  climate change and, 6–7, 9
  use of term, 6
Food Tank, 107
food waste, 191, 192–211
  amount of, 195–96
  in China, 113
  Kroger's prevention strategies, 192–98, 204–5, 207–9
  prevention strategies, 201–4
  ugly produce, 199–201, 204–5
  in U.S. cities, 198–99
forest fires, 5, 44
formaldehyde, 117
Fraley, Robb, 66, 68–70, 75
Framingham State College, 264
France, 143, 202
Franklin, Rosalind, 78
Franz, John, 102
freeze-dried foods, 2–5, 8–9, 270

freeze-drying, 8, 271
Friedberg, David, 249–50
Friends of the Earth, 72–73
Frost, Robert, 30, 51
frost candles, 51–52
Frost Dragons, 52
frost fans, 52–54
frozen fruits and vegetables, 203, 205
fruit trees, 30–56. *See also* apples
fungicides, 21, 35, 106
Future Meat Technologies, 184

G

Gandhi, Mahatma, 178–79, 192, 211
Gates, Bill, 24–25, 175, 227
Gates Foundation, 60, 62, 257–58
gene editing, 258
  CRISPR, 75, 110–11, 204, 256, 278
Generation Z, 267–69
genetically modified organisms. *See* GMOs
genetic engineering, 38, 75–76, 88, 258
Genovese, Nicholas, 173–74, 180–82, 183
Gerson, Michael, 77
Gihon, 212
Global Alliance Against Industrial Aquaculture (GAAIA), 155
"global land grab," 133–34
global warming. *See* climate change
Glover, Jerry, 250
glyphosate, 76, 94, 102–3
GMOs (genetically modified organisms), 73–79, 204. *See also* seeds of drought
  debate and controversies, 24, 60, 73–79, 88
  history of, 75–76
  "non-GMO" labels, 77
  nutrition and, 257–59
Gobekli Tepe, 14
goiter, 258
Golden Delicious apples, 76, 253
Golden Rice, 76, 257
Goldman, Josh, 155, 157, 162, 163–64
"Good-bye, and Keep Cold" (Frost), 30, 51
Goode, Erica, 156–57
Goodyear, Dana, 49
Google Ventures, 139, 140
Gore, Tim, 7
Great Veggies, 134–35
greenhouse gases, 6, 21, 46, 89–90, 250, 310–11*n*
greenhouses, 131–33
Greenpeace, 72–73
Green Revolution, 19–22, 146
  Next, 24–25
GreenWave, 164
groundnuts, 85, 249
"growth regulators," 54–55
Gulf of Mexico dead zone, 89
Gulf War, 269
Gullickson, Gil, 45

H

Hagihon, 219
Hampshire College, 163
Haramaya University, 237
Harari, Yuval Noah, 15, 290*n*
Harper, Caleb, 139, 145–46, 147, 148
Harvard Law School, 206
Harvard School of Public Health, 253, 274
HarvestPlus, 257–58
Harvest Power, Inc., 209
Harvey, Paul, 55, 57
Harwood, Ed, 129–31, 133, 134–42, 135, 143–46, 148
Haspel, Tamar, 70, 74–75
Hatfield, Jerry, 6–7, 43, 47–48, 104
Hawken, Paul, 9, 138–39, 144
Hayes, Tom, 174–75, 188–90
heart disease, 170, 174, 179, 189
"heat units," 42–43
Heimberg, Brian, 111

hemoglobin, 186
Heraud, Jorge, 9, 90–102, *92*, 104, 105–9
Herbert, Martha, 73
herbicide-resistant weeds, 97–98
herbicides, 21, 24, 89, 91–92, 95–96, 101–3, 104
heterozygosity, 34
Hjetland, Ola, 167
Hodan Fodder Cooperative, 242
Holland, greenhouses, 132–33
Holz-Schlesinger, Celeste, 187
Home Shopping Network, 4
Honeycrisp apples, 31, 37, 39, 40–41
honeydew, 114, 187
Hong, Jiang, 118–20
Hoover, Darby, 195–96, 198–201, 203–4, 211
Howard G. Buffett Foundation, 62
Humane Society, 185
hunger, 5, 11, 15, 17–18, 20–21
Hungry Harvest LLC, 205
hunter-gatherers, 15
hurricanes, 4, 5
hydroblasting, 161
hydrocooling, 52
hydroponics, 130, 131, 138, 140–41

I
IDE Technologies, 218–19, 223–24
IKEA Group, 135
Imperfect Produce, 205
Impossible Burger, 186–87
Impossible Foods, 175, 185–87, *188*
Inca Empire, 8
"In Defense of Corn" (Haspel), 70, 74–75
Indian independence movement, 178–79
Indian Tourism Association, 281–82
IndieBio, 180
indoor food production, 127–48
  Harwood and AeroFarms, 129–31, 133, 134–46, 148
  history of, 131–34
  Zhang and Tony's Farms, 111–27, 199–200
industrial aquaculture, 154–55
Inferno Limits, 136
insect farming, 251–52
insect resistance, 67, 68, 69, 71–72, 81–82
insulin treatments, 75
Intergovernmental Panel on Climate Change (IPCC), 3, 6, 7, 9
International Moringa Germplasm Collection, 243–44, 247, 248
Internet of Things, 219–20
iodine deficiencies, 258
iodized salt, 258
Iowa State University, 49
Iraq, 147, 269
Iraq War, 269
Ireland, 121, 174, 311*n*
Iron Age, 16
Ishii-Eiteman, Marcia, 83
Islamic Empire, 16–17
Islamic State in Iraq and Syria (ISIS), 17–18
Israel, water supply, 212–22
Israel Defense Forces, 215

J
Jackson, Aaron, 1–5, 8, 249–50
Jackson, Wes, 250
Japan, greenhouses, 127, 133
Jawaharlal Institute, 179
Jefferson, Thomas, 37
jellyfish, 150
jembe, 57
jewelweed, 97
Jo-Ann Fabrics, 129–30, 134
John Deere, 45, 93, 105–6, 107
Johns Hopkins University, 244
Johnson, Nathanael, 78
Jordan, 213, 221
Joseph (Genesis), 17, 25
Journal of Nutrition, 35

Juicero, 110
Julius Caesar, 51
Juniper Networks, 93
"junk food effect," 253–55
Just Meats, 184
Juul, Selina, 201–2

K
Kassa, Mitiku, 235–42, *238*
Kate & Kimi, 126
K.B. Specialty Foods, 208–9
Keillor, Garrison, 193
Kennedy, John F., 223
Kennedy, Sam, 171–72, 191
Kenya, 57–67, 72–73, 79–87, 88–89
Kenyatta University, 59
Kernza, 25, 245, 250–51, 259
Khadse, Eknath, 230–31, 234
Kibbutz Hatzerim, 217
KIND bars, 270
King Abdullah University of Science and Technology, 257
"kitchen diaries," 198
Koia, 7
Kroger, 3, 12
  Beyond Meat products, 188
  Chirps Chips, 251
  food waste prevention, 192–98, 204–5, 207–9
  Marine Harvest fish, 154
  organic foods line, 126
  PodPonics lettuce, 139
  Soylent, 273
  Zero Hunger, Zero Waste campaign, 193, 196–97
Kroger, Barney, 196
Kuli Kuli, 247–48, 256
kwashiokor, 58

L
labels, sell-by date, 193, 195, 205–7
lab meats, 173–85, 189–90
labor shortages, 55
Lambert, Mark, 223
"land grabbing," 133–34
Land Institute, 250
Lange, Christian Lous, 110
Lawrence, Ty, 172–73
leftovers, 199, 203
Leib, Emily Broad, 206
Lepeophtheirus salmonis, 150–52, 158–61
Lerøy Seafood Group, 159–60
lettuce, 71, 90–93, 117, 123, 131, 134–44
LettuceBot, 91–93, 96–97, 100, 105
lifestyle migration, 15
Liotta, Matt, 139
Liyago, Dickson, 79–83, *80*
Lobell, David, 133, 244
lobster populations, 156
Lockheed Martin Corporation, 93, 282
Loladze, Irakli, 253
Lomelde, Ingrid, 155, 157, 158, 162
London
  food waste, 202
  water networks, 216
Love the Wild, 163
Lu, Yonglong, 121–22
lumpsuckers, 159–60
lycopene, 274
Lynas, Mark, 74

M
McDonald's, 2, 174, 179, 276
McIntosh apples, 31, 41
McMullen, Rodney, 197
Madivalar, Ashok, 234
Madivalar, Honnamma, 234–35
Maharashtra, water crisis in, 230–35
maize (corn), 58, 65, 264
  Bt-maize, 68, 71, 77, 79–83
  climate change and rainfall, 47–48
  GMO seeds, 61–73. *See also* seeds of drought
  origin and cultivation, 69–70

malaria, 58, 60
malnutrition, 58, 236, 254–55
Malthus, Thomas, 18
maltodextrin, 19, 24
managed grazing, 171–72
Mann, Charles, 25–26
Manni, Alemu, 240, 241
margarine, 24
Marine Harvest, 149–55, *150*, 158–62, *165*, 165–68
marker-assisted breeding, 70, 296*n*
marred produce, 199–201, 204–5
Masters, Maxwell, 78
Matete, Mary, 84–86, *85*
Matete, Robert, 84–86
Mayet, Mariam, 66, 72
Mayo Clinic, 179
meat, 168, 169–91
  carbon footprint of, 23, 168, 171–72
  cloning cattle, 172–73
  Impossible Foods, 175, 185–87
  Memphis Meats and cultured meat, 173–85, 189–90
meat cravings, 22–23, 169–70
meat moderation, 170, 190–91
Mellsop, Gillian, 240
Melonas, Adam, 23
Memphis Meats, 173–85, 189
Mendel, Gregor, 18–19, 70
Meraas Group, 135
Mesopotamia, 14–15, 16
methane, 194, 196, 209
Michigan State University, 45–46, 49, 54–55
micrometers, 33
Microsoft, 25, 86, 215
microwave vacuum drying, 271–73, *272*
migration in China, 112
milk, 258
milk labels, 206
Mistry, Shizad, *231*, 232
MIT Media Lab's Open Agriculture Initiative, 139, 145–46
Mitsubishi, 140
Momofuku, 135, 145–46, 186–87
Mongstad, 165–66
monocrops (monocropping), 75–76, 88, 106–7
monocultures, 15, 21, 72, 163
monosodium glutamate (MSG), 24
Monsanto, 45, 59–60, 92, 96, 97
  GMO seeds, 62, 66–73, 77–83
  Roundup, 24, 76, 94, 101–2, 102–3
  Roundup Ready, 76, 77, 97–98
Moringa, 243–45, 246–49, 252, 255–59
Morrison, Denise, 267
Mosa Meat, 184
Mount Kenya, 65
Mowlid, Hamdi Muhammed, 242
Moyle, Peter, 156
MREs (Meals, Ready to Eat), 265, 267, 274
Msuya, Joyce, 7
Muhammad, 16–17
mung beans, 249
Musk, Elon, 140
Musk, Kimball, 140
Myers, Sam, 253–54
Myrothamnus flabellifolius, 84

N

"nanocrystals," 54
Nashville, 12, 192
  food waste, 198, 199, 211
National Academy of Sciences, 73
National Autonomous University of Mexico, 243
National Center for Atmospheric Research (NCAR), 232
National Environment Management Authority of Kenya, 73
National Geographic, 168
National Geographic Society, 246
National Medal of Technology, 68
National Science Foundation, 246
Natural Resources Defense Council (NRDC), 195, 211

Nature (magazine), 69, 82
nature connectedness, 11–12
Nature Conservancy, 3
Navakholo, Kenya, 60–61, 64
Neolithic farming, 14, 17
neonicotinoids, 90
Nestlé, 267
Netafim Irrigation, 218
New Age Meats, 184
"New Dust Bowl," 48
Newman, Annie, 26–29, 279–87
Newman, Chris, 26–29, *27*, 279–87
Newman, Eric, 113
New Technology Group, 84, 85
New York City, food waste, 198, 199
New Yorker, 49, 274
New York Times, 24, 156
Next Green Revolution, 24–25
Nierenberg, Danielle, 107
nitrous oxide, 89–90
Njord (god), 152
Noah's Organics, 126
nongimin, 117
NorQuin (Northern Quinoa
Production Corporation), 2, 249–50
Norwegian fjords, salmon fisheries,
149–55, 158–62, 165–66
"no-till" farming, 103–4
Nvidia, 93, 96

O

Oak Ridge National Laboratory, 84
Oaland, Øyvind, 166, 167–68
obesity, 3, 22, 88
ocean warming, 156–57
O'Conner, Lauren, 271
Octopops, 23
Odongo, Omar, 81–82
Okamoto, Michael, 260–63
Oleksyk, Lauren, 260–71, 277
OLIO (app), 202
olives, 50, 213, 215
Olson, Mark, 243, 245–47, *246*, 248,
255–57, *256*, 258–59
omega-3 fatty acids, 167, 188, 189, 244,
253
Omnivore's Dilemma, The (Pollan), 27,
171, 282–83
Oniang'o, Ruth, 57–60, *59*, 62–63,
65–66, 68, 78–79, 84, 88–89
Opal apples, 204
Orange County Sanitation District,
225–28
Orchard-Rite, 53
Organic Valley, 113, 124
Orwell, George, 260
Oshima, Marc, *135*, 135–38, 141–44
overengineering, 110–11
overfishing, 148, 155
Oxfam International, 7

P

papayas, 77
Paris Climate Conference, 186–87
Parker, Georgann, 192–95, *194*, 196,
197, 204–5, 207, 208–9
Pazazz apples, 31, 34, 39–41, 53
peaches, 43, 44, 47
Pearson, Pete, 197–98, 203–4, 205,
208, 210–11
Pederson, Byron, *231*, 231–34
Peleg, Amir, 212–22, *214*, 223–24
Pell, Willy, 96
Perfect Day, 185
"personal food computers," 145–46
personalized nutrition, 266–67
Pesticide Action Network, 83
pesticide drift, 82, 98
pesticides, 19, 21, 81–82, 88–90, 95,
98, 102, 103
agricultural runoff, 21, 89–90
on apple crops, 35, 38
in China, 112–13, 122
DDT, 24, 89
mass die-offs from use, 21
soil erosion, 102, 103, 133
photosynthesis, 48, 104, 138, 253
phytochemicals, 274

phytoremediation, 122
pigweed, 97–98, 104
Pikimachay, 14
pineapples, 76, 200
Piscataway Indians, 281, 282, 287
pistachios, 50, 52
plant-based eggs, 185
plant-breeding, 18, 25, 38, 66, 68–70, 74. *See also* GMOs
"plant factories," 127–28. *See also* indoor food production
Plenty Unlimited Inc., 127, 140
ploughs, 16, 18
PodPonics, 139
Pollan, Michael, 11, 27, 74, 75–76, 171, 282–83
pollution, 133
  agricultural runoff, 21, 89–90
  soil contamination in China, 112–13, 117–18, 120–22
polyculture, 162–63, 164
Pontifical Catholic University of Peru, 94
population, 6, 18, 19, 147
portion sizes, 203
post-food, 7, 260–78
  Soylent, 7, 261–62, 269–70, 273–77
  3-D printed food, 7, 260–63, 265–66
Potato (robot), 90–93
precision agriculture, 95, 108–9, 118–20, 123–24, 125
"preppers," 4
Prescott, Matthew, 185
presidential election of 2016, 44–45
printed food, 7, 260–63, *261*, 265–66, *266*
processed foods, 5, 254–55
propagation, 97
"Prophets" vs. "Wizards," 25–26
"protein plots," 256–57
protein sources, 14, 15, 84, 148, 154, 168, 170, 179, 241, 249, 252–53, 256–57, 273
Publix, 196
Pulp Fiction (movie), 273
Purdue University, 83
"purple pipe network," 218
purple potatoes, 253

Q
Qatar, 147
Qiwen, Qiu, 122
Queensland University of Technology, 258
quinoa, 2, 249–50, 255, 257, 258, 259

R
radiant zone drying, 271
rainfall, 46, 47–48, 65
Real Junk Food Project, 202
reaping machines, 18
recycled wastewater, 218–28
Redden, Lee, 93, 95–96, 100
Reed, Nathan, 98–102, 103–4
ReFED, 196
rehydrated pot pies, 1–2, 8–9
reinvention vs. deinvention, 25–26, 28
Research Center for Eco-Environmental Sciences, Chinese Academy of Sciences, 121
reverse osmosis (RO), 224, 225, 226, 227
Rhinehart, Rob, 261–62, 273–75
rice, 76, 162, 257
Rissler, Jane, 82
Road to Wigan Pier, The (Orwell), 260
Roberts, Paul, 19–20
Roberts, Sonya, 175–76
Robinson, Jo, 252–53
Robinson, Mary, 237
RoboCop (movie), 88
robotics, 55, 88–109, 111, 285–86
  See & Spray, 98–102, *99*, 104, 105–6, 159, 263
  Foodini, 260–63, 277
  Heraud and Blue River Technology, 9, 90–102, 104, 105–9

LettuceBot, 91–93, 96–97, 100, 105
printed food, 7, 260–63, 265–66
Sting Ray, 159–60
weeders, 90–93, 95–96, 98–102, 105–6
Rogers, James, 207
Ronald, Pamela, 74, 77–78, 83–84, 296*n*
Rosenberg, David, *135*, 135–36, 140–42
Roundup, 24, 76, 94, 101–2, 102–3
Roundup Ready, 76, 77, 97–98
Rozin, Paul, 226–27
Rural Outreach Program of Africa (ROP), 59–66, 85

S

saccharin, 24
Sacramento River, 156
Salatin, Joel, 27, 171, 282–83, 284
Salinas Valley, 90–91, 142–43
SalMar ASA, 159–60
salmon, 152, 156, 157
salmon farming, 24, 149–68
fish feed, 157, 159, 166–68
Marine Harvest and Aarskog, 149–55, 158–62, 165–68
problem of fish waste, 155, 161–62, 163, 164–65
salmonid alphavirus (SAV), 158
salmon isavirus, 158
salmon louse, 150–52, 158–61
sambhar, 247
Sam's Club, 4, 31
Sanchez, Pedro, 68
San Diego Water Authority, 222–23, 224–25
San Francisco, food waste ban, 209–10
Sapiens (Harari), 15, 290*n*
saponin, 257
Saudi Arabia, 133, 147, 224, 257
Saveur (magazine), 23
SCADA (supervisory control and data acquisition), 216
scaling, 25, 28
Scerra, Mary, 260–63
Schneider, Helene, 224–25
school intervention, 64
Schulze, Eric, 177
seafood, 154–55, 156–57. *See also* aquaculture
sea level rise, 4, 235
Sea of Galilee, 212–13, 218
seaweed, 160, 164
seed germination, 137, 141
seeds of drought, 57–87
GMO seeds and Monsanto, 62, 66–73, 77–83
history of plant-breeding, 68–70
Oniang'o and ROP, 57–66, 78–79, 88–89
WEMA program in Africa, 65–67, 71–73, 79
Seeds of Science (Lynas), 74
See & Spray, 98–102, *99*, 104, 105–6, 159, 263
self-driving cars, 94
self-steering tractors, 94–95
Seligman, Hilary, 254–55
sell-by labels, 193, 195, 205–7
sensor technology, 95, 108, 118–19, 137–38, 150, 216, 266, 278
Serralini, Gilles-Eric, 74
Shanghai, 111, 112, 115, 118–19, 123
Shanghai Tony Agriculture Development, 118
Shared Harvest, 115, 120
Shi Yan, 115, 120
Shiyuka, Michael and Amani, 61–64, 65
Sichuan Agricultural University, 117, 118
Silent Running (movie), 146–47
Silent Spring (Carson), 89
Silk Road, 34
Simple Truth, 126
Smith, Adam, 202
Smith, Bren, 164
Smithfield Foods, 113, 154

smog, 111, 112, 232
snake melons, 131, 200
snowpack, 48, 232
sodium chloride, 232
soil cleanup, 120–21
soil contamination, in China, 112–13, 117–18, 120–22
soil erosion, 102, 103, 133
soil microbes, 104–5
soil quality, in China, 112–13, 117–18
Solheim, Erik, 6
Somalia, 64, 234, 238, 241
sonic agglomeration, 270
Sonic bar, 270–71
Sorek desalination plant, 223–24, 225
soybeans, 76, 186, 249, 252–53
Soylent, 7, 261–62, 269–70, 273–77
Soylent Green (movie), 273
soy plant, 84
"specularia," 131
spirulina, 251
Square Roots, 140
Stanford University, 94, 95, 133, 187, 200, 244
Staniford, Don, 154–55, 157–58, 167
State University of New York, Buffalo, 179
Stead, Kiersten, 96
Steere, Dan, 55
Stegner, Wallace, 212
stem cells, 180
Sting Ray, 159–60, *160*
Stop Spild Af Mad (Stop Wasting Food), 201
Stuart, Tristram, *203*
"sublime," 8
subsistence farming, 57–58, 66, 256
Successful Farming, 45
sugar consumption, 254–55
Super Body Fuel, 7
Super Bowl XLVIII, 55–56
"superchilling," 42
Super Meat, 184
"superweeds," 76, 97–98, 151
survival food industry, 1–5, 8–9
sustainable agriculture and food, 11–12, 22–29, 107, 173, 284–85
Syngenta, 45, 92, 96, 97

T

Tahoe National Forest, 245
TaKaDu, 214–16, 219–20
Talpiot program, 215, 216
Target, 188, 196, 273
Technion University, 84
technology (technological fixes), 25, 110–11. *See also* GMOs; post-food; robotics
  cultured meat, 173–85, 189–90
  Ferguson and, 30–31, 33, 35, 52, 54
  Newman and, 28, 284–85
  overengineering, 110–11
  reinvention vs. deinvention, 25–26, 28
  sensors, 95, 108, 118–19, 137–38, 150, 216, 266, 278
  Zhang and Tony's Farms, 111–27, 199–200
Teeple, John, 106
teosinte, 68, 69
terroir, 143–44
Tesco, 202
Tester, Mark, 257
Tetra Tech, 198, 199
Texas drought of 2011, 48
Thailand, cloud seeding, 232
Thanksgiving, 9, 286
thermal desalination, 224
third way agriculture, 26, 79, 91, 107–8, 108, 148, 171, 211, 225, 235, 278, 281
Thoreau, Henry David, 34
3-D printed food, 7, 260–63, *261*, 265–66, *266*
Tiberius, 131, 133, 200
Tien Tsin, 115
tilapia, 155, 168
tilling, 16, 103–4

toilet-to-tap programs, 225–28
Tony's Farm, 113–27, *116*, 118, *125*
Tony's Spicy Kitchen, 114–15, 117–18
Too Good to Go, 202
tools, 15–16, 276–77
Toshiba, 139
trade, 16–17
Trader Joe's, 77
trans fats, 24
Trimble, 94–95, 96
truffula tree, 244, 258
Trump, Donald, 44–45
2,4 dichlorophenoxyacetic acid (2,4-D), 97–98, 102
Tyson Foods, 2, 154, 171, 174–75, 188–89, 310*n*

U

ugali, 58, *72*
Ugly Mugs, 205
ugly produce, 199–201, *200*, 204–5
"underchilling," 42, 47
United Arab Emirates (UAE), 147
United Nations, 65, 147, 155, 179, 221
  Environment Programme (UNEP), 6, 7
  Food and Agriculture Organization (FAO), 240–41, 242
  World Food Programme (WFP), 236, 240
Unitywater, 220
University of California, Davis, 74, 156
University of California, San Francisco, 254
University of California, Santa Barbara, 245
University of Cape Town, 84
University of Florida, 47, 68
University of Maryland, 282
University of Minnesota, 180
University of Nottingham, 248–49
University of Wisconsin, 31, 37, 43
"updraft," 232
urbanization, 112
USAID (United States Agency for International Development), 62, 236
USDA (United States Department of Agriculture), 6–7, 43, 47, 77, 102, 104
  Climate Change Program, 48–49

V

Valentine's Day Peach Massacre, 43
Valeti, Uma, 173–85, *177*, *184*, 189–90
vaping, 175
vata, 169–70
vegan burgers, 185–88
vegetable box schemes, 23, 115–16
vegetarianism, 170–71, 178, 179
vegetarian salmon feed, 166–68
Veneklasen, Gregg, 172
vertical farms, 127–28, 133, 135–42
  Harwood and AeroFarms, 129–31, 133, 134–46, 148
Vietnam War, 89
vitamin A, 76
vitamin C, 71, 266, 272
vitamin D, 258
Vitamin Shoppes, 251
viticulture, 50, 51

W

Wageningen University, 237
Wallace, Alfred Russel, 57
Wall Street Journal, 78, 186
Walmart, 4, 31, 150, 188, 196, 273
Walsh, Margaret, 48–49
Walt Disney World, 251
Wang, Jay, 122
Wang, Rose, 251–52
Washington Post, 70, 74–75, 77
Washington State University, 54, 59
Waste Management, 192–93
wasting food. *See* food waste

water conservation, 217
water crisis, in Maharashtra, 230–35
Water Efficient Maize for Africa (WEMA), 62, 65–67, 71–73, 79
water filtration, 225–28
water leaks, 214, 215–16, 217, 219–20
water pricing, 221–22
water quality, 218, 222
Waters, Alice, 11, 173
water scarcity, 82–83
water supply, 48, 212–28
  in California, 221–28
  in Israel, 212–22
  recycled wastewater, 218–28
water theft, 220
Watson, James, 78
Weather Modification, 230–35
weeding robots, 90–93, 95–96, 98–102, 105–6
Wefood, 201–2
Wegmans, 131
West Bank, 213–14
West Texas A&M, 172
wheat, 69
  Kernza, 25, 245, 250–51, 259
White Castle, 186, 188
Whole Foods, 111, 115, 116, 150, 163, 196, 197, 205, 248, 273
wild apples, 34
wild fisheries, 155, 156–57
wild salmon, 152, 156, 158
Willett, Walter, 274
Wilmoth, Gabriel, 96
Wilson, Bee, 23
Wisconsin Apple Growers Association, 33
Wisconsin freeze of 2012, 31, 45–46, 52
Wisconsin freeze of 2016, 30–31, 30–33, *32*, 36–37, 40, 42–43, 44–45, 52–53
Wise Company, 1–5, 8
  Chicken-Flavored Pot Pie, 1–5, 8–9
WISErg, 209
"Wizards" vs. "Prophets," 25–26
Wolffia globosa, 249
World Coffee Research, 51
World Food Prize, 68
World Food Programme (WFP), 236, 240
World Health Organization (WHO), 24, 73, 76, 103, 112
World War II, 8, 102
World Wildlife Fund (WWF), 155, 197, 203–4, 208
wrasse, 159–60

Y
YaData, 215
Yang, Tom, 271–73
Yang, Xiaohan, 84
Yinon, Zohar, 219–20

Z
Zarchin, Alexander, 218
Zero Hunger, Zero Waste campaign, 193, 196–97
Zestar apples, 31, 34
Zhang, Tony, 111–25, *114*, 126–27, 199–200
Zhang, Xiaoming, 126
Zhao, Findle, 124
Zhou Dynasty, 162
Zhu, Jian-Kang, 83
Zoninsein, Manuela, 111

ABOUT THE AUTHOR

AMANDA LITTLE is the author of *Power Trip: Fro*[illegible]*lls to Solar Cells – Our Ride to the Renewable Future*. An aw[illegible]g environmental journalist, she has written for the *Ne*[illegible]*s*, *Vanity Fair*, *Wired*, *Rolling Stone* and the *Washingto*[illegible]g others. She teaches investigative journalism at Vander[illegible]ty and lives with her husband and children in Nashville, Te[illegible]